Physics and the Rise of Scientific Research
in Canada

Until the First World War, universities in Canada were dedicated primarily to liberal education and professors invested most of their energy in teaching and the preparation of courses. Since then, universities have been transformed to serve a new generation of scholars devoted to research. Yves Gingras explores this dramatic shift through a study of the development of the physics community in Canada.

The teaching of engineering and a change in liberal arts curricula, both stimulated by industrial growth, encouraged the creation of specialized courses in the sciences. By the 1890s, Gingras argues, trained researchers had begun to appear in Canadian universities. The technological demands of the First World War and the founding, in 1916, of the National Research Council of Canada accelerated the growth of scientific research. The *Transactions* of the Royal Society of Canada could no longer publish everything they received because of the disproportionately large number of research papers from the society's scientific sections. In response NRC created the *Canadian Journal of Research*, dedicated to the publication of scientific research. By 1930, a stable national system of scientific research was in place in Canada.

With the dramatic increase in the national importance of their disciplines, scientists faced the problem of social identity. Gingras demonstrates that in the case of physics this took the form of a conflict between those who promoted a professional orientation, necessary to compete successfully with engineers in the labour market, and those, mainly in the universities, who were concerned with the problems of the discipline such as publication, internal management, and awards.

This is the first book to provide a general analysis of the origins of scientific research in Canadian universities. Gingras proposes a sociological model of the formation of scientific disciplines, distinguishing the profession from the discipline, two notions often confused by historians and sociologists of science.

Yves Gingras is a professor in the Département d'histoire, Université du Québec à Montréal.

Physics and the Rise of Scientific Research in Canada

YVES GINGRAS

TRANSLATED BY
PETER KEATING

McGill-Queen's University Press
Montreal & Kingston • London • Buffalo

© McGill-Queen's University Press 1991
ISBN 0-7735-0823-6

Legal deposit first quarter 1991
Bibliothèque nationale du Québec

Printed in Canada on acid-free paper

This book has been published with the help of a grant
from the Social Science Federation of Canada, using
funds provided by the Social Sciences and Humanities
Research Council of Canada. Publication has also been
aided by a grant from the Comité des publications,
Université de Québec à Montréal.

This book is a translation of *Les origines de la recherche
scientifique au Canada*, © Boreal 1991. Translation has
been assisted by a grant from the Canada Council.

Canadian Cataloguing in Publication Data

Gingras, Yves, 1954–
  Physics and the rise of scientific research in Canada
  Translation of: Les origines de la recherche scientifique
  au Canada
  ISBN 0-7735-0823-6
  1. Physics – Canada – History.
  2. Physics – Research – Canada.
  I. Title.
  QC9.C3G5513 1991   530'.072071   C90-090514-X

Typeset in 10 on 12 Baskerville by
Nancy Poirier Typesetting Ltd.,
Ottawa, Ontario.

To the memory of
my mother, Liliane,
and my father, Jean-Marie

Contents

List of Figures and Tables

Acknowledgments

Research drawn out over the course of many years is possible only with the collaboration of many individuals. I wish to thank particularly Othmar Keel, who read several versions of this text and made numerous useful comments which bore as often on content as on form. Marcel Fournier has given me constant encouragement in my research and has offered constructive criticism on the numerous texts I have given to him over the past six years. My thanks to him. Richard A. Jarrell, who has done so much for the history of science in Canada, has always supported my research, and he kindly read two preliminary versions of this book. I thank him for his generous and constructive comments.

I have also benefited from a stimulating intellectual environment thanks to numerous exchanges with Raymond Duchesne, Robert Gagnon, Peter Keating, Creutzer Mathurin, Michel Trépanier, and, during my visit at Harvard University, Silvan S. Schweber. I also appreciated George Weisz's support while preparing this book. Thanks to Paul Dufour, Phillip Enros, Craig Fraser, and Donald Phillipson, who, in Ottawa and Toronto, unhesitatingly sought out documents needed for this book, thus saving me much travel. I wish to thank my research assistant, Denis Veilleux, who diligently reverified a number of facts and helped me to collect new information. Thanks, finally, to Henry Small, who kindly furnished me with information concerning citations to physics journals.

Historians cannot work without archivists. Without conservation and classification, a large number of documents would not be available to researchers. I would like to thank Mario Creet, of the Queen's University Archives, and Charles Armour, of Dalhousie University, who gave me precious help in the course of my research.

Like much university research, the present work has been facilitated by scholarships awarded by the Fonds FCAR of the Quebec government and by the Social Sciences and Humanities Research Council of Canada. I would like to thank those in charge of these organizations as well as the evaluators who, between 1979 and 1984, found my project worthy of support.

Parts of this book have been previously published in a different form in *Critical Issues in the History of Canadian Science, Technology and Medicine*, Richard A. Jarrell and A. Roos, eds. (Ottawa: HSTC Publications, 1983), 16–30; *Scientia Canadensis*, 10 no. 1 (May 1986), 53–71; *Canadian Historical Review*, 68 no. 2 (1986), 181–94; *Histoire sociale/Social History*, 19 no. 37 (May 1986), 73–91; and *Youth, University and Canadian Society: Essays in the Social History of Higher Education*, Paul Axelrod and John Reid, eds. (Montreal: McGill-Queen's University Press, 1989), 301–19. Thanks to the editors for permission to reuse this material here.

The final acknowledgment must go to those who come first in my life: Sylvie Duchesne and our daughter, Dominique.

Physics and the Rise of Scientific Research in Canada

Introduction

The status accorded research in the present-day university may give the impression that, along with teaching, the production of new knowledge has always been one of the essential functions of this institution. However, the conception of the modern university as both a research and a teaching institution is relatively recent. Until the beginning of the nineteenth century, the university was, generally speaking, a teaching institution which trained "men of character" in law, medicine, and theology. A professor was defined above all as a teacher and devoted himself to his courses and his students; the qualities valued were eloquence, erudition and paedagogical excellence. A typical professor prepared a textbook as the natural outcome of his work and hoped thereby to profit, in symbolic terms at least, from a lifelong investment in course preparation by influencing future generations.

With the creation of the University of Berlin in 1810, a completely different type of professor appeared. The reform of the Prussian universities, inaugurated by Wilhelm von Humboldt, minister of education from 1809 to 1810, redefined the role of the professor by obliging him to be first and foremost a researcher. At the time, scientific researchers were usually found either in scientific academies or in their own private laboratories. Any association with a university was generally in a teaching capacity. The researcher defined his obligations principally in relation to the scientific community to which he belonged and within which he circulated the results of his research. Restricting himself to a particular problem, the researcher was essentially a "specialist," and the result of his activity was an article published in a scientific journal. Recognition came neither from students nor from the university but from other researchers, his "peers."

Creation of the University of Berlin thus provided this class of individuals with a unique site for institutional reproduction. Whereas the "natural philosophers" of the seventeenth and eighteenth centuries were the somewhat unique products of improbable social trajectories, those of the nineteenth century were more standardized and came from the university. This beginning of the institutionalization of the production of researchers helped accelerate the growth and specialization of knowledge.

We are generally well informed about the institutionalization of scientific research in countries with a well-established scientific tradition – such as Germany, France and England – as well as in those with a more recent tradition of scientific research – such as Australia, Japan, and the United States; but we know little about how research activities took root in Canadian universities.[1] This book seeks to fill that gap by presenting a study of the formation of the Canadian physics community. In effect, this group of researchers was the first scientific community to be constituted in Canada. As we shall see, its spokesmen, while drawing support from researchers in other disciplines, played a central role in the setting up of structures which, between the two world wars, stimulated research in most scientific disciplines.

At a more general level, this study offers an analytical model for the study of the formation of scientific communities. Most studies to date rely on the impressionistic use of such concepts as "professionalization" or "disciplinarization," which are rarely, if ever, precisely defined. Such theoretical weaknesses have resulted in histories that relate events without being able to show the relationship between these events and a set of predefined questions that give meaning to the narrative.[2]

Since scientific research is produced by social agents with particular characteristics, a description of the emergence of this activity supposes explaining the production of this type of agent and of its relation with the scientific field. In other words, in preparing a genealogy of the group of Canadian physicists (or, indeed, any other group of scientists), one must first show how the transformation of the educational system, conceived of as an "apparatus for the production of agents,"[3] made possible the emergence of these agents, by instituting a new form of pedagogical action which inculcates the scientific "habitus" – a system that generates practices, perceptions, and evaluations of practice.[4]

I therefore seek to describe, in an almost ethnographic fashion, the trajectories of a new generation of scientists who perceived themselves in terms more of their research activities than of their teaching duties. The centre of analysis thus becomes the process of disciplinary socialization, which transforms the scholarly characteristics of the new professors. Unless we believe in the spontaneous conversion of agents to the practice of research, we must conceive of a "production apparatus" that can form agents to function in this particular type of practice.[5]

Sociologists have understood the importance of the socialization process. As Warren O. Hagstrom has written: "The socialization of scientists tends to produce persons who are so strongly committed to the central values of science that they unthinkingly accept them. Research as an activity comes to be 'natural' for them ... These commitments are the outcome of a prolonged training process, lasting well into adult life, in which the student is effectively isolated from competing vocational and intellectual interests and in which he is extremely dependent on his teachers."[6]

This process of inculcation takes place according to a dynamic appropriate to scientific disciplines. We may, following Bourdieu, conceptualize these disciplines as fields that are sites of "a competition in which the stake is the monopoly of scientific authority inseparably defined as technical capacity and social power."[7] Now, "in order for a field to work, there must be stakes and people ready to play the game. The players must be endowed with the habitus which implies the knowledge and recognition of the immanent laws of the game, the stakes."[8] Material resources and institutions must be available so that the agents may produce the knowledge that circulates in the field. From this point of view, which emphasizes the practice of research as opposed to discourses on this practice, a national community in a scientific field may be conceived of as emerging in three distinct phases.

The first phase consists of the emergence of a research practice. In a country with a young scientific tradition, such as Canada or Australia, this usually happens through importation of know-how via apprenticeship in a European laboratory. In countries with an older scientific tradition, the emergence of a new practice may be explained through reconstruction of individual trajectories which have yet to be institutionalized. Thus, the emergence of organic chemistry around Justus von Liebig at Giessen in the 1840s and the work of the first German electrodynamics researchers in the 1810s are examples of new practices which became institutionalized because of favourable historical conjunctures.[9]

Once conditions for emergence of the practice of research have been brought together, the first representatives of this new category of agents can impose a conception of the university institution which is compatible with their research activities and which allows the long-term reproduction of the group. This second phase, the institutionalization of research, is crucial to the formation of all national scientific communities. It is, in effect, the condition for the growth and survival of a community, for only by setting up institutional structures which favour the production of knowledge and the reproduction of agents endowed with the necessary dispositions for this activity can scientists reproduce themselves as a group and participate in activities that constitute a scientific field.

Finally, the formation of a social identity – either disciplinary (i.e. creation of a scientific association) or professional (i.e. creation of a corporation) – represents the third phase in the formation of a scientific community. By giving themselves official representatives, scientists acquire a social visibility and can thus defend their interests (and those of their discipline) by propagating a particular image of themselves and their discipline.

Such a model can guide historical inquiry insofar as it allows one to problematize phenomena often taken for granted. It is, of course, a relatively general analytical framework: its content varies from one country to another according to the course of events, and it presupposes no evolutionary trends. Given the diversity of necessarily singular historical experiences, this model can highlight common traits and thus perhaps help explain why the organizational forms which constitute scientific disciplines are, after all, similar in most countries.[10]

Applied to Canada, the model allows one to link together transformations in sectors that have traditionally been studied separately by historians: for example, the emergence of engineering education, the modification of education programs, the creation of diploma and bursary programs in higher education, the formation and evolution of the Royal Society of Canada and the National Research Council of Canada, and the role of funding agencies and scientific journals in the research process. All these changes must be taken into account if we are to understand the formation of scientific disciplines within the universities.

This study may appear only descriptive. However the model just outlined structures the narrative: it serves as a guide for the descriptions which are constructed as explanations of the transformations that occurred during the period studied. These transformations made possible the formation of the scientific communities, which are reproduced in the university. "Theoretical" terms, which often merely repeat in other words what the narrative has already said, may mask the actions of social actors who, through their institutional and cognitive interests, create historical change. In order to be understood, these actions must be restored to the real-life context of multiple possibilities and it is the function of narrative to reconstruct these contexts and the movements of the actors.[11]

In the case of Canadian physics, the formation of a scientific community took a bit more than a century, from approximately 1850 to the 1960s. The three major elements in this process overlap chronologically and are reflected in the three parts of this book – "Establishing Research" (chapters 1–3, covering roughly from 1850 to 1920), "Reforming

Institutions" (chapters 4 and 5, from 1900 to 1950), and "Changing Definitions" (chapters 6–8, from the middle of the First World War to the mid-1960s).

Chapter 1 describes the conditions under which there emerged a first generation of Canadian physicists in the period 1850–1900 and shows how its research-oriented practice differed from that of the previous generation, which had taught in Canadian universities and whose activity was essentially pedagogical. This chapter brings to light the indirect link between the development of engineering education and the growth of science departments. It also shows that reform of the bachelor's program and creation of specialized "honours" sections during the 1860s allowed scientists to specialize in a particular discipline, thus creating a possible trajectory for future researchers.

In chapter 2, we shall see that the first researchers who held posts in the university, had, in order to develop as a group, to work to institutionalize their practice by promoting creation of structures that facilitated production of knowledge and the reproduction of agents endowed with the necessary dispositions. Although the discourse on the importance of research in the "modern" university began in the 1870s and multiplied from the beginning of the twentieth century, only in the 1910s did war and the problems of Canadian industry give nationwide coverage to these claims, hitherto confined within the walls of the universities which sheltered active researchers.

As we shall see in chapter 3, the institutionalization of research in physics brought about rapid growth in the production of scientific publications and in the number of graduates in master's and doctoral programs and diversification of research.

In chapter 4, we shall see how the problem of communication between researchers and adequate diffusion of their results in the international scientific field led researchers to transform the Royal Society of Canada and the only existing national scientific publication, the *Transactions* of the Royal Society of Canada. Founded in 1882, the journal was inadequate to the dynamic of the scientific field, where agents can acquire credibility only by assuring themselves priority for their discoveries. Researchers in the universities had to modify the *Transactions* in order to accelerate its publication and increase its diffusion, so that it could play an effective role in the field of international science. However, their efforts did not succeed in completely transforming the publication.

Chapter 5 will show how the National Research Council of Canada, created in 1916, intervened in 1929 to found the *Canadian Journal of Research*, the first scientific journal in this country set up specifically for researchers.

Physicists still had to decide how to organize themselves and how to present themselves to the world. In chapters 6, 7, and 8, we shall see how and in what circumstances Canadian physicists, between the First World War and the 1950s, endowed themselves with organs for representation and spokesmen charged with defending the interests of the group and of the discipline and with propagating a "public image" of the physicist which assures the group's future by attracting the recruits necessary for its reproduction. This analysis will enable us to see how the notion of "professionalization" – used so often by historians and sociologists of science to describe the evolution of scientific disciplines – corresponds to nothing more than one of several means of forming a group identity.

In sum, through the formation of a scientific discipline – physics – we shall see in this study the emergence of Canada's first system of scientific research.

PART ONE

Establishing Research

CHAPTER ONE

From Teaching to Research

During the second half of the nineteenth century, physics teaching in Canadian universities underwent important changes. From "natural philosophy," which was taught to all students in arts faculties and which required only the services of a professor trained according to the traditional bachelor of arts model, we move to the establishment of physics departments located in premises which, although not always entirely devoted to the purposes of this discipline, had sufficient space for the installation of laboratories where students could take momentary leave of their textbooks and be introduced to the techniques of experimental physics.

The major cause of this transformation was the mid-century demand for engineering education created by Canada's industrial development. These needs induced a modernization of the university curriculum, heretofore dominated by philosophy and classics. The institutionalization of engineering education, as well as the development of medical faculties during the 1870s, thus created a growing demand for science courses, which thus assured rapid development of physics departments and departments of science in general from the 1880s on.

The early 1870s saw also the transformation of physics professors. Trained in Europe at a time when laboratory experience had become an integral part of scientific training, the new physics professors tended to invest as much time in research as in course preparation. They thus behaved less as "professors" than as "researchers." At the turn of the century – by the time this new generation had, with great difficulty, trained disciples – the first cohort of researchers, mostly physicists, trained in Canadian universities made their appearance.

ECONOMIC DEVELOPMENT AND
UNIVERSITY REFORM

In the early 1850s, the traditional structure of the major Canadian anglophone universities, set up in the first decades of the century along English and Scottish lines, began to be called into question by the economic and industrial activity brought about by the first boom in railway construction.[1] The massive and sudden presence of this new technology could not help but generate discussion about the importance of the sciences in the modern world in Quebec, Ontario, and the Maritime colonies.

The first signs of a reform movement in Canadian higher education appeared at more or less the same time in Quebec, Ontario, and the Maritime provinces. In Montreal, a report prepared in 1844 on the courses offered at McGill College indicated that the structure of the Arts Faculty, with its traditional teaching program, no longer met the needs of a commercial city like Montreal. The report recommended introduction of more modern courses such as moral philosophy, political economy, civil engineering, and agricultural chemistry.[2]

This new conception of education brought about the modification of McGill's charter and the hiring of William Dawson as university principal in 1855. His mandate was to adapt the institution to the needs of Montreal's anglophone community.[3]

Dawson had been Nova Scotia's superintendent of education from 1850 to 1853 and had helped modernize that province's school system by introducing such subjects as natural history, agricultural chemistry, and even some elements of physiology into the secondary-school curriculum.[4]

In New Brunswick, Dawson also helped transform King's College into a provincial lay university. The lieutenant-governor, Sir Edmund Walker Head, had advanced the idea of modernizing the courses offered at King's College in Fredericton. A committee of the governing body, which contained partisans of Head's idea, subsequently recommended "that more specific attention be given Civil Engineering, that is, to its leading principles."[5] A first series of conferences was presented at the beginning of 1854 by the English engineer MacMahon Cregan.

However, the college was not prepared to introduce this type of teaching, and Head formed a commission to inquire into the situation. Modelled after the 1850 Oxford Commission of Inquiry in which Sir Edmund had participated, the commissioners included Egerton Ryerson and William Dawson, the latter invited by the lieutenant-governor to participate in early 1854.[6] Tabling their report in the legislative assembly several months later, the commissioners suggested that the government create a provincial lay university offering courses not only in civil engineering, but also in agricultural chemistry, commerce, and navigation, subjects previously absent from King's College, whose curriculum and management were more adapted to a pre-industrial society.[7]

These recommendations, brought into effect in New Brunswick only in 1859, with the creation of the University of New Brunswick, were put into practice almost immediately at McGill by Dawson. In his inaugural speech in 1855, Dawson explained his vision of university teaching. Having recalled the importance of traditional subjects, he turned to the sciences: "Independently of their charms and value, [sciences] have in our day established a connection so intimate with every department of mechanical, manufacturing and agricultural art, that without them the material welfare of nations cannot be sustained, much less advanced."[8]

Perhaps taking advantage of the recent opening of the Toronto-Montreal railway, he invited the engineer Thomas Coltrin Keefer, author of *The Philosophy of Railroads*, to give a series of talks on civil engineering and railroads in the fall of 1856. These talks launched the program in civil engineering announced formally in the university calendar the following year. Spread out over two years, the program led to a diploma in civil engineering.[9] In Ontario, the situation evolved in a similar fashion. In 1850, King's College was replaced by the University of Toronto, created by the provincial government, and from 1857 a diploma in civil engineering was offered.[10]

These first attempts at engineering education were, however, a false start. The stop in railway construction affected student recruitment: at McGill, the program was suspended in 1863 for lack of students and because of a series of financial difficulties. Nonetheless, McGill did manage to produce fifteen graduates in the 1860s. In New Brunswick, the number of students remained low, and, at Toronto, there were only four graduates between 1859 and 1870.[11]

This modernization movement also had an impact in French Canada where the increasing demand throughout the 1860s for a more practical education led to the creation of the Ecole Polytechnique by Quebec's provincial government in 1873. However, the Polytechnique, marginalized in an educational system dominated by the clergy, recruited few students and had no real influence on the development of the sciences in other francophone institutions. Given what occurred in Canada's anglophone universities, it is likely that the Polytechnique's status of independent corporation hindered the development of scientific education at the Université de Montréal. The growth of science departments elsewhere resulted in large part from the demand generated by engineers and physicians, but professors of chemistry, physics, and biology at the Université de Montréal were deprived of an engineering clientele and were forced to rely on the Faculty of Medicine alone.[12]

THE INSTITUTIONALIZATION OF ENGINEERING
EDUCATION AND THE GROWTH OF SCIENCE
DEPARTMENTS

Despite the eloquent discourses of the representatives of "modernity," the educational structures were linked most significantly, if in a complex and indirect fashion, to the economic market. This connection is clearly demonstrated by the suspension of the civil engineering program at McGill and the stagnation in enrolment in Toronto and Fredericton in the 1860s. The institutionalization of engineering education means more than offering a few courses. It supposes the creation of institutional structures that provide positions for professors, space for laboratories, technicians to take care of laboratory instruments, all of which demand considerable investment and, of course, a steady student clientele.

It was only with the revival of railway construction in the 1870s and the start of a number of public works programs that a more stable demand for civil and mining engineers facilitated the emergence of viable schools, faculties, and departments of engineering in Canadian universities.[13] Announcing the reopening of the engineering program at McGill in May 1871, Dawson noted: "The times appear to be particularly favourable, in consequence of the present activity in mines, railways and other scientific enterprises in this country."[14]

As the level of industrialization was higher in Quebec and Ontario than in the east, these were the first provinces to have structured engineering education. Moreover, as Montreal and Toronto were more populous than other Canadian cities, McGill and the University of Toronto were in a position – both economic and demographic – to dominate among Canadian universities for a long time.[15] At the University of New Brunswick, for example, enrolment grew slowly: it was not until 1927 that an engineering department distinct from the physics department was founded. Queen's University opened a School of Mines in 1893 but was in constant struggle with the University of Toronto for both government subsidies and students.[16]

As Robin Harris has noted,[17] the development of science departments is directly linked to the demands of engineering and medical schools, and it was predictable that science departments would develop first at McGill and the University of Toronto. For this to happen, required that the schools of medicine and engineering be affiliated with the universities.

This was the case at McGill, but at Toronto such an affiliation was contested. Thus, during discussions over the creation of a College of Technology by the Ontario government in 1873, there was opposition to its proposed affiliation with the University of Toronto: "science teaching had not been a success, the University having failed to meet the scientific needs of the country."[18] Conscious of the importance of this affiliation for the future of the university, its directors were forced to intervene in order to assure governmental support. James Loudon, professor of physics at the University of Toronto, later wrote:

To my mind, this period was the most critical one in the history of the University, for had the Government of that day lived to establish a strong and efficient College of Science to teach not only the higher technical subjects, but the physical and Natural Sciences as well, the cause of Science in this University would probably have received a fatal blow, and the Arts Faculty would have ultimately been reduced to the status of a small College. Fortunately for all the educational interests concerned, there was a change of Government ... and among the new ministers there were some who [would] countenance no step which was calculated to weaken the Provincial University.[19]

With allies in the government, the university finally won. Named professor of mathematics and natural philosophy in 1875, Loudon was soon called on to advise the government on the development of the College of Technology. Unsurprisingly, Loudon recommended affiliation to the university in order to allow: "The utilizing of the Science professoriate of University College for the instruction of Technical as well as Arts students. This plan ... involved the establishment of four laboratories for the Department of Chemistry, Physics, Mineralogy and Geology and Biology, all of which were provided in 1878."[20]

While the number of students enrolled in science courses during the period 1860–78 had remained stable, following affiliation it grew steadily.[21] In 1887, the student body had grown sufficiently to justify the separation of physics from mathematics. Loudon therefore left the latter discipline to his assistant, Alfred Baker. A departmental structure similar to that found in American universities began to emerge.[22]

At McGill, as we have seen, the engineering course was revived in 1871. Here again, growth in enrolment was rapid, and in 1890, three years after Toronto, a chair in physics was created and John Cox, a Cambridge graduate, became the first holder.[23]

Though often linked to the growth of engineering schools, a chair in science may arise under other circumstances. Benefiting from a grant from the businessman George Monro, Dalhousie University was able to create Canada's first chair in physics without any subsequent growth in the department. It was only after creation of a chair in civil engineering in 1903 and establishment of the Nova Scotia Technical College in 1909 that the physics department was able to hire additional professors.[24]

Although at the turn of the century science departments were well-defined institutional spaces within Canadian universities, their principal function remained the servicing of the professional faculties. Thus, in welcoming Alexander Duff in 1890 to the chair in physics, the president of the University of New Brunswick recalled: "The professor of Physics has two practical aims in view; firstly to give students of civil engineering that acquaintance with general physical principles without which the highest success in their chosen profession is impossible; secondly, to assist student who aim at Electrical Engineering as a profession."[25]

Three years later, when John Cox inaugurated the Macdonald Physics Building at McGill – on his own admission, an "unforeseen" development in a plan that had called originally for construction only of the Engineering Building – he recalled that the primary purpose was the training of future engineers and doctors. Moreover, the laboratory was also to respond to the needs of the Arts Faculty: "A knowledge of physics is now-a-days a necessity in any liberal education, and no curriculum which calls itself liberal can afford to ignore the study of the laws of nature and the methods of the science of investigating them."[26]

This insistence on the importance of physics for arts students reminds us that while the growth of engineering education carried science education along with it, this causal relation is not eternal. Thus, at the University of Toronto at the turn of the century, the growth of student enrolment in the Arts faculty was such that the physics department, which in the mid-1870s had expended so much energy to attract students from the School of Applied Sciences, was able to survive the 1904 separation between the school and the university, despite Loudon's efforts to prevent it.[27] As we shall now see, the arts faculties contained the true future of physics in Canada.

REFORM OF THE BA PROGRAM AND
THE BEGINNINGS OF SPECIALIZATION

As we have just seen, the emergence in the mid-1850s of a utilitarian conception of the sciences in Canadian universities, stimulated by economic development, first materialized in specific programs for the applied sciences. A second transformation, equally important for the future of physics, led to a change in the bachelor of arts program, opening new career possibilities for students who, before 1860, had but one path before them and no possibility for specialization in any of the sciences.

The old program had been based on "the principle of encouraging a well balanced and varied culture, and not with the view of stimulating extraordinary proficiency in particular departments,"[28] and had sought to educate individuals who would later pursue the traditional careers of law, medicine, and theology.[29] Natural philosophy was presented as a collection of general knowledge that all students should acquire, but of secondary importance in comparison to philosophy and the classics. In the final years of study, the student was introduced to the basic principles of dynamics, hydrostatics, optics, and astronomy.[30]

The introduction in 1860 of more specialized ("honours") courses in the BA program, modified this traditional organization. As ten years before, with the emergence of engineering education, this new attack on the dominant conception of teaching met with resistance from those who considered the change likely to lead to undue specialization.[31] Nonetheless, the pragmatic vision of education which we found in the discourses on the importance of the applied sciences also imposed itself here and led to the diversification of the courses offered to BA students.[32]

While the development of engineering education made possible the emergence of physics departments by supplying clients, the modification of BA studies put into place the elements necessary for the institutional reproduction of physicists. At the University of Toronto, for example, in 1882, the fourth year of "honours" studies was divided into two options: physics and mathematics. The student who chose the physics section could then "spend more time in the laboratory and, in some cases, participate in research projects."[33] Three years later, experimental work in the laboratory, which had been instituted in 1878 on an optional basis, became mandatory for all students in the "honours" section.

The emergence of these new paths of study marks the beginning of the slow institutionalization of physics in Canada. Before this official recognition of the possibility of a physics *career*, the physicist had no institutional basis for disciplinary reproduction: all students followed a homogeneous program of mathematics and physics, where the latter subject did not have the status that would have allowed it to attract a particular clientele.

As we shall now see, science professors who taught in nineteenth-century Canadian universities were also the products of a system based on the classical humanities. Their training produced not "physicists" or "scientists" as we know them today but, rather, people prepared more for teaching in general than for any science in particular.

DECLINE OF A PEDAGOGICAL PRACTICE

Most professors of natural philosophy teaching in Canadian anglophone universities in the nineteenth century had been trained in English and Scottish universities prior to the reform of these institutions that followed the new wave of industrialization beginning in the 1870s.[34] Despite differences vis-à-vis English universities such as Oxford and Cambridge, Scottish universities inculcated the same taste for general culture, the only legitimate culture in the eyes of the traditional élites. As a former Glasgow student recalled: "The Arts curriculum of that day was intended for culture only, not as providing something which was to have commercial value; the idea that any of its subjects could be made the foundation of industrial training in science as now understood and that the students could specialize in such a subject had not then been contemplated."[35]

Called on to teach in an institution which, although located thousands of kilometres from their alma mater, was only a copy of the one that had trained them, the graduates of Glasgow, Edinburgh, or Cambridge were therefore perfectly adapted to reproducing the culture that the colony sought to import.[36]

Thus the holders of such titles as "Professor of Mathematics and Natural Philosophy" were not much different from their fellow professors of philosophy or Graeco-Latin literature who had followed essentially the same trajectory. William B. Jack, who taught mathematics and physics at King's College in New Brunswick between 1840 and 1885, had to teach Latin and Greek when the holder of this chair died in 1870. Similarly, before succeeding him, Thomas Harrison had taught English and philosophy.[37] In Ontario, at Queen's College, Kingston, James Williamson also taught rhetoric, encouraging his students to write essays on the nature of the solar system or on the difference between physics and chemistry.[38]

Thus, the practice of this generation of professors was more pedagogical than scientific, and few among them produced a large number of publications. In fact, the most active were at Toronto, where they attended the meetings of the Canadian Institute. Founded by a handful of engineers and surveyors in 1849, the Canadian Institute was quickly monopolized by University of Toronto professors when many became members in 1852.[39] That year, the institute began to publish the *Canadian Journal*, which reprinted papers presented by members at the society. Like most provincial learned societies of the time, most of the work presented dealt with the natural sciences (botany, zoology, archeology, and so on). Of sixty papers published in astronomy between 1852 and 1890, two-thirds were concerned with meteorological observations, most of them provided by the successive directors of the Toronto Meteorological and Magnetic Observatory.[40] There were only twenty publications in physics – most from professors at the University of Toronto, and half of these from James Loudon.

It was before this local society, whose only rallying point was an interest in the sciences, that the professors of mathematics and natural philosophy at the University of Toronto, J.B. Cherriman and, later, James Loudon, presented their work. Their participation in the activities of this learned society allows us to characterize more closely the relationship the professors maintained with knowledge. This relationship seems to have been a direct extension of their pedagogical practice.

Professor at the University of Toronto from 1853 to 1875, John Bradford Cherriman had studied at St John's College, Cambridge, where he was sixth wrangler in the Mathematical Tripos of 1845.[41] Coming to Toronto as associate professor, for a number of years he directed the Meteorological and Magnetic Observatory which had been affiliated with the university since 1853. As director, he regularly discussed the meteorological situation of the city before members of the institute, before leaving his post to G.T. Kingston.[42]

As a product of Cambridge, Cherriman was inclined toward mathematics and made many presentations on algebra and geometry which have no sense of unity and which demonstrate a rather eclectic interest in the sciences: he often presented work he had come across in the *Quarterly Mathematical Journal* and the *Philosophical Magazine* or reflections on problems that had arisen in the course of his teaching.

As an example, and though published after he had left the university to become dominion insurance superintendent, his article "The Motion of a Chain on a Fixed Plane Curve," presented at the first meeting of the Royal Society of Canada in May 1882, illustrates this pedagogical practice. Stipulating, first of all, that "the chain being supposed inextensible, the velocity of it at a given instant must be the same, and therefore so also is the acceleration," Cherriman went on to solve the general equation and then apply it to particular cases.[43] The argument was similar to that found in textbooks of the time. No reference was made to prior work, and the problem existed only in the context of an apprenticeship type of situation, where the potential student would be called on to familiarize himself with Newton's laws and differential calculus.

His successor as professor, James Loudon, reproduced essentially the same type of practice. The first Canadian-born person to become a professor at the University of Toronto, he received his bachelor of arts from that institution in 1862. After some time as tutor under the direction of John McCaul, professor of Greek and Latin, he became Cherriman's assistant.[44] He was elected member of the Canadian Institute in 1870 and presented many papers. Of seven articles (mean length: three pages) published in the *Canadian Journal* between 1871 and 1882, six dealt with problems in mechanics and one with the properties of a straight line. As with Cherriman, the work originated in the course of reflection on pedagogy. "On Trilinear Coordinates," published in 1871, begins as follows: "The following method of treating the problem of the straight line occurred to me in 1867, since which time I have used it with advantage in the lecture room."[45] In "Notes on Mechanics," published in November 1873, he posed a series of problems concerning the combination of forces. The originality of his solutions was only that they differed from those offered in traditional textbooks. The same procedure was used in his "Notes on Statics."[46]

At McGill University, the activities of physics professors did not differ from those of their colleagues in Toronto. Between 1857 and 1890, physics and mathematics at McGill were taught by Alexander Johnson. Born in Ireland in 1830, he graduated from Trinity College, Dublin, in 1854. Johnson became dean of the Arts Faculty and vice-principal of the university.[47] The Royal Society of Canada, of which he was a founding member, gave Johnson the opportunity to play an active role in the preparation of the meeting of the British Association for the Advancement of Science held in Montreal in 1884. In collaboration with the Greenwich Royal Observatory, he participated in the preparation of the Canadian program for the observation of the transit of Venus, 6 December 1884.[48]

His constant representations to the dominion government led to creation of the Survey of Tides and Currents in 1893, under the direction of the Department of Fisheries.[49] In physics, he produced only one article, of an essentially pedagogical nature, on Newton's work in optics.[50] By 1890, the growth in enrolment at McGill justified a greater division of labour, and Johnson henceforth devoted himself to mathematics.

A new professor, John Cox, was hired to teach physics. Professor of physics from 1890 to 1909, Cox had not received any research training.[51] Graduating from Cambridge in 1874, he had ranked eighth in the Mathematical Tripos and had also obtained good marks in the Classical Tripos. Before leaving England at the age of thirty-nine, he had been director of Cavendish College for ten years. Elected member of the Royal Society of Canada in 1892, he presented only two papers before the society: one written in collaboration with Callendar, on the properties of x-rays, and a description of an experiment conducted in a class he gave on the formation of frazil during which he believed he observed ice falling below the water surface.[52] Having described the various stages of the experiment, he concluded: "I have no explanation to offer, and in spite of the witness of members of the class and my own vivid recollection, now finds some difficulty in believing that the ice really sank to the bottom." Cox never returned to the subject.

If, instead of defining a generation in terms of an arbitrary chronological period of twenty-five or thirty years, as is often the case, we define it in terms of institutional stability, such that agents produced by the institution during a given period are endowed with essentially the same dispositions, because they are submitted to the same conditions of reproduction, then John Cox belongs to the same generation of professors as Cherriman, even though the latter was trained twenty-five years earlier.[53] Cox, Cherriman, and Loudon's papers are characteristic of the essentially pedagogical practice of this generation of professors – a practice distinct from the research practice of the generation trained after 1875 – which was inculcated by institutions preparing them to reproduce, not produce, knowledge. Moreover, their institutional situation forced them to draw on their course material as a source for their presentations to Canadian Institute or the more selective Royal Society of Canada. These two aspects – the agent's habitus and institutional position – were mutually reinforcing, for one must accompany the other in order for any institution to function harmoniously.[54]

As we shall see later, problems arose when a new generation of agents, more adept at scientific research, emerged at the end of the nineteenth century. Institutions had to adjust to respond to the needs of these specialists. It was only after a major modification of university structures, which allowed "research" into the organizational schemes, that a fruitful relationship with the institution could be pursued by agents whose ideal was the training no longer simply of "honest men" but of specialists who would join the ranks of scientists and contribute to the "advancement of knowledge."

Waiting for these "mutants," who would modify the knowledge relationship in Canadian universities, the professors continued their work as educators. As their pedagogical thought-schemes were best realized in the production of textbooks – which allowed them some symbolic profit at least from their investment in course preparation, which constituted the essence of the work of a good professor – Cherriman, Cox, and Loudon also published physics and algebra manuals.[55]

FRENCH CANADA: PRIEST-EDUCATORS AND
CLASSICAL CULTURE

The transformations in the English-Canadian educational system which were outlined briefly in the preceding sections (and which served as the basis for the development of physics in Canada) did not occur in French Canada. There the educational system, organized around the classical colleges and controlled completely by the clergy, retained for a long time its primary function of reproducing of the traditional élites and their culture.[56]

Founded in 1852, Laval University prepared its students for law, medicine and theology. Science teaching was elementary; it was intended above all to familiarize students with different aspects of the knowledge of nature. French Canada was consolidating a program centred on the humanities, while the new anglophone Canadian élites were criticizing this ideal of a classical training, as represented by "Oxbridge," as no longer fitting to the needs of an industrializing society.[57]

The new Catholic university, a property of the Quebec Seminary, did not have the people to fill the chairs of physics, chemistry, astronomy, mathematics, and natural history that they had decided to create. Although obliged to call on the services of laymen (such as the doctor Hubert Larue, of the medical faculty, and the American Thomas Sterry Hunt, for chemistry), the directors sought to recruit professors from among the clergy, despite the latter's rather elementary scientific training.[58]

This recruitment policy made each new science professor a carbon copy of his predecessor, of whom he was more often than not a student. The clerical monopoly on education, which the economic and social situation of Quebec held in place until the First World War, permitted the mimetic reproduction of science professors (in physics, in particular) to perpetuate itself until 1920, making real development in science education impossible. In order to see this process at work and to see how it accounts for the absence of scientific careers for francophones in Quebec before 1920, I shall describe the "itineraries" of the three successive professors of physics at Laval University who held the position from 1859 to 1925.

The first professor of physics at Laval, Thomas-Etienne Hamel, born in Quebec City in 1830, studied at the Seminary in that city and entered the Grand Séminaire in 1849 in order to pursue studies in theology in preparation for the priesthood, which he entered in 1854.[59] During this period he carried out a number of teaching assignments in the Petit Séminaire. He was simultaneously master of discipline, professor of language, physics, and mathematics, and assistant director of students. These duties were the lot of theology students recruited by the directors of the Seminary. Once finished his theological education, Hamel undertook further training in order to prepare himself to teach the sciences in the new university. From 1854 to 1858, he studied at the Ecole des Carmes in Paris, where he became a licentiate in science; he returned to Quebec ready to assume his new duties as university professor. He was assigned not only physics, which he taught until 1874, but also astronomy (until 1866) and mineralogy and geology (until 1870). As it was a matter merely of presenting the basic elements of these sciences and not at all of training "specialists," Hamel no doubt had few problems. In fact, his pupils were mainly philosophy students from the Quebec Seminary, to whom were added students in medicine who had not finished their classical courses.

Given the clerical monopoly over the French-Canadian system of education, the trajectories of the Laval professors (at least those who were members of the clergy) unfolded in a space linked inseparably to the religious field: a promotion in one more often than not led to an advancement in the other. In 1871, Hamel was named superior of the Quebec Seminary and by virtue of this became rector of Laval University; he was also named general vicar of the archdiocese. Fifteen years later, he was elected president of the Royal Society of Canada and Pope Leo XII named him protonotary apostolic.

His successor, Joseph-Clovis Kemner Laflamme, followed more or less the same trajectory. After having completed his studies at the Petit Séminaire where he was Hamel's pupil, he entered the Grand Séminaire in 1868 and was ordained in 1872 at the age of twenty-three. From 1870, he was responsible for courses in geology and mineralogy, and five years later he inherited the physics course, which he did not give up until 1893, when he became rector of the university. The same year, he was named protonotary apostolic, and the following year he obtained the title monsignor.[60]

The parallels we have looked at are supposed to suggest not a causal relationship between the two types of position (and titles) but rather that we cannot artificially separate the "university," "religious," and "scientific" careers of these figures. In fact, I believe that the term "priest-educator" (drawn from Hamel's biography) better characterizes the totality of the careers of Laval science professors than "scientist," because their training (classical and theological) could not help but turn them into more or less enlightened educators, depending on the state of the scientific discipline to which they devoted themselves.

Moreover, their position within the "Canadian scientific community" – which in the early 1880s sought an identity through creation of the Royal Society of Canada and its annual meeting – was tied more to their place in Quebec society than to their scientific work.

When the governor-general, the Marquis of Lorne, was preparing in 1881 to found the Royal Society of Canada, he had to secure the participation of French Canadians for obvious political reasons.[61] In literature and the human sciences, the solution was creation of two sections, one francophone (section I) and one anglophone (section II). However, as the sciences were considered international, a linguistic division was not conceivable, and francophone and anglophone scientists were forced to belong to the same section. Professor of physics from 1859 to 1875 and rector of Laval from 1871 to 1880, Hamel represented a politically logical choice for section III (chemistry, physics, and mathematics); his colleague Laflamme (professor of physics from 1875 to 1893 but mainly professor of geology), who had just published a college textbook on mineralogy, geology, and botany, was a natural choice for section IV (geology and biology).

Their nominations owed little to the importance of their scientific publications. Laflamme's preceded the true beginning of his work in geology, which he started in the summer of 1882 for the Geological Survey of Canada. This work gave him material for ten papers, which he presented at the annual meetings of the Royal Society between 1883 and 1907. His geological practice – essentially the collection of facts that he subsequently sent to specialists – bear the mark of an incomplete theoretical training.[62] Unable himself to identify fossils, a central procedure of the geological practice of the time, he could not actively participate in discussions of the discipline's problems. His interest in these debates was expressed only in the publication of popularizations, which revealed his talents as a teacher. In 1891, in Le Canada-français, an intellectual journal of general interest published by the professors at Laval, he published an article on the controversy surrounding the geological identity of Quebec. In the scientific field, the debate had taken place, without Laflamme's participation, in scientific journals such as Science or the American Journal of Science.[63]

While Laflamme made minor contributions to geology, Hamel contributed nothing to physics. His rare presentations to the Royal Society of Canada were more exercises in rhetoric typical of classical training. The title and even the structure of his papers bear the mark of the rhetorician's culture. Presented in 1884, his "Essai sur la constitution atomique" was full of questions and exclamations but made no reference to scientific work. In "De la certitude dans les sciences de l'observation," presented in 1891, Hamel attacked evolutionary ideas, opposing them to the "bonne méthode scientifique inductive."[64]

Notwithstanding individual variations, the intellectual (and social) trajectories of Hamel, Laflamme, and their successor, Henri Simard, who taught physics from 1893 on, resembled those of abbé Ovide Brunet, professor of botany at Laval until 1870, and abbé Léon Provancher, botanist and entomologist and founder of the Naturaliste canadien, who died in 1892.[65]

In becoming "priest-educators" or country priests (as Provancher and Brunet had been for a while), these products of the classical colleges followed a path determined by the educational program of these institutions: the reproduction of the clergy and the liberal professions. Achieving a particular position, even marginal, in the scientific field (as opposed to an amateur practice which, by definition, is outside the field) depended not only on self-education but also on the structure of the disciplines they chose to practise.

If, by the mid-nineteenth century, botany and entomology still needed the empirical work of specimen collection and classification,[66] allowing Brunet and Provancher to do useful work in botany and entomology, the situation was more complex in geology, which, at the time Laflamme began work, was in full transformation. The development of theories and of knowledge at the end of the century necessitated greater specialization and increased use of techniques and precision instruments, which for all practical purposes excluded self-taught amateurs such as Laflamme.[67]

If Laflamme's access to geology constitutes a limiting case, Hamel's to physics is not: he had no means of integrating himself into the clan of "physicists," even in the marginal fashion of Loudon or Cherriman, who, using the resources of their pedagogical practice, could at least talk about physics.

Reproducing the itinerary of his predecessors, Henri Simard, who taught physics at Laval from 1893 to the mid-1920s, was still further excluded. While Hamel elected to one of the scientific sections of the Royal Society of Canada, the development of research from 1900 on and the appearance of the first generation of Canadian physicists definitively excluded "amateurs," and Simard entered as a member of section i. Elected in 1923 by literary scholars, this priest-educator had played his role as pedagogue by publishing physics and astronomy manuals for use in colleges as well as producing commentaries on science which he read before public audiences at Laval.[68]

If francophone scientists were somewhat rare at the turn of the century, it was not because the possible trajectories were modified but because the few doors to which they led were closed: the era of talented amateurs, which had allowed Laflamme, Brunet, and Provancher to contribute to the advancement of science was over.

In French Canada, the educational system tended to reproduce an essentially cultural relationship to science. The logic of reproduction, which entails use of the products of the institution to reproduce the same culture, explains the extreme rigidity of the system. This rigidity made it difficult if not impossible to produce a new relationship to science. When, following the First World War, industrialization forced the modernization of higher education on university authorities, it was necessary to import, from Switzerland, the first professors for the Ecole supérieure de Chimie. It was the only way to break the circle of reproduction. In physics, the scenario was the same. The first two directors of the physics department at Laval were Italian physicists: Franco Rasetti, from 1939 to 1947, followed by his friend Enrico Persico, who held the position until 1950. The Université de Montréal, which gained independence from Laval in 1920, followed the same model and, in 1945, hired a French physicist, Marcel Rouault, to modernize its physics teaching and to stimulate research.[69]

As the first generation of francophone physicists was trained only in the early 1950s, when the discipline was already well established in other

Canadian universities, the emergence of the social group took place essentially within anglophone universities.

Between the generation of natural philosophy professors who taught in English Canada before 1875 and that which succeded it in the last quarter of the century, a change in the position of the sciences in English and Scottish universities altered the gamut of possible trajectories and thus allowed emergence of a true specialization in physics, which in turn made possible the production and reproduction of physicists.

Incarnated in the Oxford and Cambridge Commission of Enquiry set up in 1850, the reform movement brought about the creation of a Natural Science Tripos at Cambridge and more regular science teaching. The development of engineering education and the success of the second transatlantic telegraph cable in 1866, as well as the perception of English industry as performing poorly (particularly in comparison to German) at the Universal Exposition in Paris in 1867, led to official recognition of physics laboratories and their incorporation into teaching programs.[70]

Between 1866 and 1874, ten physics laboratories appeared in Cambridge, Edinburgh, Glasgow, and London, opening the doors to a new type of teaching, where purely formal training gave way to an apprenticeship centred more and more on the laboratory.[71] For physicists such as William Thomson or Peter Guthrie Tait, for example, who until then had received just a few students in private laboratories, it now became possible to inculcate an entire generation of students with dispositions regarding experimental research that they themselves had acquired in a more haphazard manner (i.e. outside a particular institutional framework). This transformation, which institutionalized the social characteristics of a new type of agent, was quite similar to that which gave birth to the Liebig School of organic chemistry. Indeed, as J.B. Morrell has shown, Liebig's laboratory "was the first institutional laboratory as opposed to private laboratory in which students experienced systematic preparation for chemical research, and in which they were deliberately groomed for membership of a highly effective research school."[72] After having passed through these new centres of apprenticeship in Britain and similar ones in Germany, the Canadians and British called on to teach in Canada implanted a research practice in its universities.

Born in Green Hill in Nova Scotia in 1847, John James Mackenzie obtained from Dalhousie a BA in 1869 and an MA in 1872. He then decided to continue his studies in Europe and enrolled at Leipzig University, where he obtained his doctorate in 1876. Working with

Gustav Wiedemann, who since 1871 had occupied the first chair in physical chemistry in Germany, he wrote a thesis on the absorption of carbonic acid by saline solutions. The results were published in Wiedemann's journal, *Annalen der Physik und Chimie*.[73] Mackenzie then worked for some time in Berlin under the direction of Helmholtz and produced "a number of fine and difficult experiments on the matter of electromotive force induced by magnetism in insulators," as Helmholtz remarked in his letter of recommendation to Dalhousie University.[74] Replacing his friend J.G.MacGregor, who had just obtained a better position in England at Clifton College in Bristol, he was hired as lecturer in the fall of 1877.[75] He died prematurely, in February 1879, having published only two more articles in Wiedemann's journal. The first dealt with Kerr's theory of the relation between optics and electricity, and the second with the variation in the volume of a solution following absorption of a gas. This article was written in collaboration with the American physicist E.L. Nichols, whom he had met in Wiedemann's laboratory.

Faced with financial problems, the Dalhousie University authorities managed to convince George Monro, a rich New York businessman born in Pictou, to endow a chair bearing his name with an annual salary of $2,000. This new chair enabled the university to offer satisfactory teaching in physics.[76] The George Monro Chair of Physics, the first of its kind in Canada, was offered to J.G. MacGregor, who decided to return to his alma mater, having left two years earlier unsatisfied by the salary of $750 a year.

MacGregor's academic training was similar to that of his friend and predecessor, Mackenzie.[77] After having obtained his BA from Dalhousie in 1871 at the age of nineteen, he received a Gilchrist Scholarship[78] which enabled him to pursue his studies at the University of Edinburgh, where he worked in Peter Guthrie Tait's laboratory. In 1873, he published the results of his first research in the *Transactions* of the Royal Society of Edinburgh.[79] This work, on the electrical conductivity of saline solutions, led him naturally to Wiedemann's laboratory in Leipzig, where he joined his friend Mackenzie and carried out experiments on electrolytic resistance. On the basis of this work, he obtained his doctorate from the University of London in 1876.

When he returned to Dalhousie to take over the first chair in physics, MacGregor had already published seven papers in the *Proceedings* and *Transactions* of the Royal Society of Edinburgh, five of which bore on the thermoelectric properties of metals, in a period (1860-80) when physicists were greatly concerned with the electrical conductivity of metals and solutions.[80]

Isolated from the international scientific community, MacGregor was pleased to participate in the founding of the Royal Society of Canada and used its first meeting, in May 1882, to present his work on the electrical conductivity of solutions. Read at the same session as Cherriman's paper on the movement of a chain in a plane, MacGregor's highlighted the distance that separated these two generations of university professors trained in different educational systems. "On the Measurement of the Resistance of Electrolytes by Means of a Wheatstone Bridge" began with a presentation of current research, emphasizing unsolved problems:

> The chief difficulty in the measurement of the resistance of electrolytes is due to the Polarization by the current of the electrodes of the electrolytic cell ... Special precautions must be taken either to reduce or to remove the Polarization ... The bridge method of Kohlrausch, Nippold and Grotrian was based upon the reduction of polarization ... Professor J.A. Ewing and myself found it possible to make ... determinations of the resistance of electrolytes ... by simply using a galvanometer.[81]

While Cherriman and Loudon's papers were self-contained, having only to give the ideal conditions of the problem in order to expose the solution, MacGregor's paper was essentially open: he showed how the problem he had chosen had been treated by others, as well as the relation of his work to theirs. In a word, MacGregor's work was part of a research program – a system of relations obtaining between a totality of work which circulates essentially in the scientific field – while his predecessors' was the product of pedagogic reflection. Where MacGregor referred to peers and to the laboratory, Cherriman and Loudon referred to the textbook.

While occasionally addressing pedagogical issues,[82] MacGregor concentrated on problems of physical chemistry. In the beginning, he even used his summer vacations to "make use [through the kindness of Professor Tait] of the rich stores of the Natural Philosophy Laboratory of the University of Edinburgh."[83]

After having applied, without success, for the position of professor of physics at McGill, he managed in 1890 to put together a small laboratory at Dalhousie. As he wrote in 1901: "Following the traditions of the Edinburgh Laboratory, I have endeavoured to stimulate my students to engage in research. My Advanced Practical Class was organized for this purpose eight years ago, and during this time a number of investigations have been made which have given results worthy of publication."[84]

MacGregor directed the work of at least eight pupils who published a total of seventeen papers, all devoted to the physico-chemical properties of aqueous solutions. Although these papers were published in a local journal, the *Transactions of the Nova Scotia Institute of Science*, three also appeared in the *Philosophical Magazine*. This introduction to research enabled several of MacGregor's pupils to obtain scholarships and to pursue doctoral studies in the United States.

I have already mentioned that John Cox, when he arrived at McGill in 1890, was assigned to direct construction of the Physics Building, which, when it opened in 1893, was one of the best equipped in the world.[85] The same year, the generosity of William Macdonald, the tobacco magnate who gave millions of dollars to McGill,[86] enabled the university to hire a second Macdonald Professor of Physics. Another Cambridge product, Hugh Longbourne Callendar, then crossed the Atlantic. Unlike Cox, however, Callendar belonged to the new generation of "researchers." Trained at the Cavendish Laboratory under the direction of J.J. Thomson, Callendar was a good example of the transition from classical to scientific trajectory made possible by the institutionalization of physics laboratories in the late 1860s.

In his autobiography, Thomson describes with great detail how he "transformed" Callendar, whose classical training had ill-prepared him for the laboratory, into an excellent experimental physicist:

H.L. Callendar's career at the Laboratory was in some respects the most interesting in all my experience. He was on the classical side when at school and did not any physics. As an undergraduate he took a First Class in classics in 1884 and one in mathematics in June 1885. He came to work in the Laboratory in the Michaelmas Term in that Year. *He had never done any practical work in physics, nor read any of the theory* except in a very casual way. He had not been in the Laboratory for more than a few weeks when I saw that he possessed to an exceptional degree some of the qualifications which make for success in experimental research ... *The problem was to find a subject for his research which would give full play to his strong points and minimise as much as possible his lack of experience* ... It seemed to me that the most suitable research would be one which centred on the accurate measurement of electrical resistance.[87]

Callendar's school trajectory was thus one of transition. Trained in a humanist tradition, which prepared him in no way for physics, he was able, under Thomson's direction, to modify his trajectory, while minimizing the disadvantages of his initial training. As the honours science courses and science programs developed, preparing students for direct entry into the laboratory, this type of reconversion became less and less necessary.

Like his Dalhousie colleague, whom he soon met at meetings of the Royal Society of Canada, Callendar arrived at McGill with a research program. Having succeeded in making a platinum resistance thermometer, he had hopes of using it to make precise thermodynamic measurements. Luckily for Callendar, high-precision work based on electrical measures matched McGill's laboratory equipment, which included an excellent collection of batteries and resistors calibrated according to international standards.[88] In addition to having precision instruments, Callendar was able to count on the collaboration of two demonstrators hired in 1894 to help in the preparation of laboratory experiments which had to be executed by close to two hundred students.[89]

Thus Henry Marshall Tory (BA, 1894) was called on to study the thermoelectric properties of iron using Callendar's platinum thermometer in a project which, according to Cox's report to the university's president, had "also an important bearing on some experiments on cylinder condensation which were carried out ... by Nicholson and Callendar at the Thermodynamic Laboratory of the Macdonald Engineering Building."[90] His fellow student F.H. Pitcher (BASC, 1894) observed variations in the magnetic susceptibility of metals as a function of temperature. Another applied sciences graduate, Howard Turner Barnes, joined the team of demonstrators in 1895, and Callendar busied him with research into the variation in the potential of Clark's cells. These researches formed the basis of the first master's theses in physics at McGill.[91]

After having proved himself in the colony, Callendar returned to the metropolis, where, in 1898, he received a position at University College in London.[92] At this juncture, another of Thomson's students, Ernest Rutherford, took the post in Canada, introducing a research topic destined for success: radioactivity.[93]

Although not able to create a school – as he had intended – like his master Thomson, Rutherford none the less trained a dozen students in his nine years in Montreal.[94] As we shall see, some of them found positions in Canadian universities and became part of a group of physicists whose task it was to construct a solid basis for the discipline.

A later founder of the discipline of physics in Canada, John Cunningham McLennan, originated what might be called the "Toronto School." Born in Ontario in 1867, he entered university after having taught primary school for several years. A product of the physics section inaugurated in 1882, McLennan graduated from the University of Toronto in 1892. Loudon soon hired him as demonstrator in physics. In the summer of 1896, McLennan travelled to Europe and visited such well-known physics laboratories as Warburg's, where Röntgen had just discovered x-rays. Having decided to do research, he wrote in the 1896–7 university calendar that "special arrangements may be made by graduate students for pursuing original investigations in the laboratory."[95]

Conscious of the limits of his training, he took a year's leave from the university and went in 1898 to the Cavendish Laboratory (just when Rutherford had left for McGill), where he worked under Thomson on the electrical conductivity of gases submitted to cathode-ray bombardment. Published in the *Proceedings of the Royal Society of London*, this work earned him, in 1900, the first doctorate in physics granted by the University of Toronto, the degree having been instituted only three years earlier. Following in Rutherford's footsteps, McLennan quickly surrounded himself with several assistants and attempted in vain to outdistance his McGill rival. Always abreast of the latest developments in physics, McLennan directed an important research team in Toronto, which, beginning in 1910, specialized in spectroscopy and remained, until the Second World War, the centre of physics in Canada.[96]

PROBLEMS OF REPRODUCTION

It should not be thought that the two dozen students introduced to research between 1893 and 1907 by Callendar, MacGregor, and Rutherford (McLennan did not begin training students until 1910) all followed the new path and devoted their lives to research. If learning the trade constitutes the first stage in the reproduction of the researcher's habitus, one must next find a position that allows one to do research. At the turn of the century, only the university brought together the necessary conditions by offering, in additon to time and money, a few instruments and sometimes an assistant, which allowed professors to undertake research.

Moreover, not all these students had the same training, and so, depending on whether or not they were engineers, the range of employment offered them was more or less varied. At McGill, for example, all the students working in the laboratory prior to the creation of the BSC program in 1898[97] came from the Faculty of Applied Sciences – except for H.M. Tory, who had a BA – and were trained in electrical engineering. For research, this training was perfectly adequate for the essentially electrical manipulations undertaken in Callendar's lab. However, as few positions were available in physics and as industry more willingly hired engineers, Callendar's students did not spend much time in the universe of research and usually found employment as engineers.

Having obtained a master's degree in 1897 and having earned a living for several years as a demonstrator, F.H. Pitcher followed a career as an engineer at the Montreal Water and Power Co. His colleague R.O. King took advantage of the 1851 Exhibition Scholarship (a bursary which, as we shall see in chapter 2, was expressly conceived, in 1890, to train researchers) in 1895 and within several years was directing the King Construction Co. in New York. In 1909, this position enabled him to offer a $600 bursary to young researchers in the physics department.[98] R.W. Stovell, Callendar and H.T. Barnes's assistant, found work at Westinghouse in New York, and J.L.W. Gill, following King as the 1851 Exhibition Scholar in 1897, received a position in electrical engineering at Queen's University.

Of all of Callendar's students, only H.T. Barnes, who became associate professor in 1901 after having subsisted for five years as a research assistant, would have the time and the institutional resources to continue a research program begun in 1895 with his mentor's impetus. His colleague H.M. Tory, who in 1903 was professor of mathematics at McGill, devoted forty hours a week to course preparation, which left him little time for research. Tory's career as a physicist was therefore quite short, and he devoted the rest of his life to education, founding several universities across Canada. In 1927, at the age of sixty-three, he became president of the National Research Council of Canada. As we shall see in chapter 5, he played an important role in the creation of the *Canadian Journal of Research*, the first national journal to respond to the needs of Canadian researchers.

While Rutherford insisted on the "pure science" aspect of his work and considered Callendar an engineer, his colleague, Barnes, reproduced this applied physics practice and thus oriented his students toward industry rather than university. Thus, R.G. Duncan, a 1903 graduate in mining engineering, published several papers in collaboration with Barnes and left the laboratory to work for the Grand Trunk Pacific Railway. Before 1910, Barnes directed four theses. The authors all came from the Faculty of Applied Science, and most of them found work in industry.

Recruiting his students (including one woman) from the Arts Faculty, Rutherford had a better rate of institutional reproduction than Barnes, as many of his students obtained university positions and thus achieved a certain level of scientific productivity, which, though not always highly visible, assured them an existence in the field of physics. After graduating bachelor of arts in 1899, Robert Kenneth McClung worked on the rate of ion recombination as a function of pressure. He received his master's degree in 1901 and the 1851 Exhibition Scholarship, which took him to Cambridge for three years of work under Thomson's direction. He spent the three following years at McGill as a demonstrator and then became professor at Mount Allison University in New Brunswick. Taking advantage of the 1851 Exhibition Scholarship, both H.L. Cooke (BA, 1900; MA, 1903) and Robert William Boyle (BSC, 1905, MSC, 1906) followed the same trajectory: the former studied at Cambridge from 1903 to 1906 and then became professor of physics at Princeton; the latter went to Manchester and then returned to McGill as associate professor. Rutherford's first female student, Harriet Brooks (BA, 1898,

MA, 1901) began a promising career at Bryn Mawr and at Cambridge, where Thomson ranked her as one of his most able collaborators. She left research to marry F.H. Pitcher, one of Callendar's former students, and lived in the anonymity of family life.[99] Throughout their careers, McClung, Cooke, and Boyle remained in contact with Rutherford, even though their institutional positions did not allow them to follow the path laid down for them by their mentor.[100]

At Dalhousie University, the only other possible site for the training of physicists at the end of the nineteenth century, we have seen that MacGregor had trained eight students between 1893 and 1900. Dalhousie did not train engineers at the time but offered a BSc diploma, which was MacGregor's source of students. His work on electrolytic conductivity and ionic theory was situated at the crossroads of physics and chemistry. Consequently, some of his students became chemists, while others became physicists.

His first student, Frederick James A. McKittrick, published a paper on electrolytic resistance in 1894 and received, the same year, the 1851 Exhibition Scholarship to study at Cornell University in the United States. He worked for two years constructing a Hotchkiss' galvanometer and measuring the energy dissipation in an iron ring submitted to an alternating magnetic field. He did not obtain a degree, and in 1905 he became managing director of the General Electric Co. of New York. More productive than McKittrick, E.H. Archibald, an 1897 graduate, published five papers in the two following years and received an MSc in 1898. With his colleague Douglas McIntosh, he was one of the most productive chemists of the time.[101] Archibald's work on the conductivity, density, and surface tension of aqueous solutions earned him in 1898 the 1851 Exhibition Scholarship, which enabled him to finish his doctorate in physical chemistry at Harvard University. Returning to Canada, he joined McIntosh as a demonstrator in the chemistry department at McGill.

McIntosh obtained his BSc from Dalhousie in 1896, and his work with MacGregor led him, too, to publish a paper on the calculation of the conductivity of an electrolytic mix. Named 1851 Exhibition Scholar in 1896, he studied at Cornell and Leipzig before becoming demonstrator at McGill in 1901. Elected fellow of the Royal Society of Canada in 1909, he rose in the university hierarchy to become full professor of chemistry in 1914. His first papers, on the relationship between the solubility and freezing point of compound solutions, as well as those on the electro-motive force produced by a gas battery, led him to direct theses in collaboration with Barnes in the physics department. Thus it was that the first two research programs developed in Canada by MacGregor and Callendar came together in George W. Shearer's thesis, submitted to the physics department in 1908, on the electromotive force produced by the contact of magnesium and aluminum in the presence of various electrolytes. The following year, Percy H. Elliott deposited a thesis directed by McIntosh and Barnes on the conductivity of certain salts in solution.[102]

MacGregor's first students, James Barnes, T.C. Hebb, and T.C. Mackay, went into physics. The first two studied respectively at Johns Hopkins and Chicago, thanks, once again, to the 1851 Exhibition Scholarship, and obtained their PHDs in 1904. Mackay paid his own way to Harvard, where he received his doctorate in 1903. Following similar trajectories, they all received positions as professors of physics late in the First World War.

Between 1893 and 1908, MacGregor, Callendar, and Rutherford's first attempts to train researchers resulted in the initiation of twenty young Canadians to research. Half of them – all students of Callendar, except for Barnes – found work in industry and ceased to participate in the production of knowledge through publications in physics. Those who found a position in a Canadian university were therefore disciples of MacGregor (T.C. Hebb and D. McIntosh) and Rutherford (R.K. McClung and R.W. Boyle). Along with McLennan and several American physicists, these young researchers attempted to consolidate their institutional position after 1910.

The gestation of the first generation of physicists trained in Canadian universities was carried out in rather artisanal conditions. Even at McGill, where Rutherford made fundamental discoveries, only William Macdonald's generosity allowed the university to overcome the organizational deficiencies of an institution that had not been designed for research.

At the turn of the century, Canadian universities were teaching institutions, not research centres. Professors were naturally hired to teach, and, in terms of time, their research activities were marginal. However, the most important aspect of the transformations we have studied was not that the researchers taught less than their predecessors but rather that they did not define themselves primarily in terms of their teaching duties and conceived of their role in terms of research. In a sense, we can say that their attention was now directed toward the laboratory, not toward the lecture hall. It is this passage from a pedagogical practice to a research practice that, in my opinion, led to a modification of the traditional discourses on the "mission" of the university and to the establishment, at the turn of the century, of master's and doctoral programs.

The emergence of scientific research, in physics at least, at the end of the nineteenth century was therefore not the effect of a new social demand or the needs of Canadian industry but was rather the indirect consequence of changes which, in Europe, had affected university science teaching during the 1860s and the 1870s, by giving the laboratory an important role as a training site. However, the infiltration of this new generation of research-oriented professors into Canadian universities was facilitated by the growth in science teaching generated by the development of engineering education, which responded directly to the needs of the industrializing nation.

In order to get out of their marginality, though, the researchers had to give wider diffusion to their discourses on the importance of scientific research and thus go outside the walls of the university. As we shall see in the next chapter, it was only during the 1910s that the theme of the importance of scientific research acquired a national dimension and contributed to the consolidation of the researchers' position within Canadian universities.

CHAPTER TWO

Finding Resources

In the preceding chapter, we followed the scholarly trajectories of science professors in the principal Canadian universities so as to observe the emergence of a new generation of scientists who, trained in Europe, imported a new practice into Canadian institutions. It is now necessary to take a closer look at what may be termed the "material conditions of production," which allowed first H.L. Callendar, Ernest Rutherford, and J.G. MacGregor and subsequently their successors to work as researchers. At the end of the nineteenth century, there was no room in the university structure for "research," which the new heirs wished to introduce. A certain production of knowledge – and reproduction of researchers – had been possible, but in the absence of favourable institutional conditions. There had been little financial support and laboratory space, the university degrees were ill-adapted to this new mode of reproduction, and there was little hope of employment for these new "researchers," who were and still are dependent on the university milieu for the pursuit of their "disinterested" practice.

After having reviewed the origins of the first bursary programs which allowed many young Canadians to acquire research training in Europe, we shall see how, on their return, these people were forced constantly to demand resources (in terms of time, space, and money) in order to continue their research projects. Until the First World War, their demands remained those of agents isolated within their institutions. It was not until the National Research Council of Canada (NRC) was founded in 1916 that their discourse on the importance of research was taken up by institutions and acquired greater legitimacy, since it was now declared to be in the national interest.[1] A true research system, with bursary programs for students and grants to researchers, then appeared and consolidated the position of research in physics and other scientific disciplines whose development until then had rested with a few researchers who had to make do with the resources at hand.

Generally speaking, the trajectories followed by the first generation of Canadian researchers point to the importance of scholarships in these researchers' training. Thus, J.G. MacGregor at Dalhousie and A.W. Duff at New Brunswick were able to study in Europe thanks to the Gilchrist Scholarship, and, beginning in 1891, many science students at Dalhousie, McGill, Queen's and Toronto used the 1851 Exhibition Scholarship to study in the United States and Europe. Historians have noted the importance of these scholarships for the development of higher education. According to Hugh Hawkins, "the fellowship as an award to attract graduate students ... was probably the crucial institutional invention that brought success to the early Johns Hopkins."[2]

In Canada, before the First World War, there was no true bursary program that favoured higher education. Moreover, as doctoral programs had appeared only at the beginning of the twentieth century, students wishing to be trained as researchers were forced to leave the country at their own expense or win a bursary offered by a foreign university, in particular, in the United States.

Prior to establishment of the large number of scholarships offered from the 1880s on by American universities, Canadian students were able to participate in the Gilchrist Scholarships.

FORERUNNER:
THE GILCHRIST SCHOLARSHIP

The Gilchrist Scholarship, inaugurated in 1868 and applicable to any discipline, was awarded annually to a Canadian who wished to study for a BA degree at either the University of London or the University of Edinburgh.[3] In practice, however, most holders of the scholarship had already received their first degrees in Canada and saw little benefit in further undergraduate study. Most would use the scholarship to study at the post-graduate level. This implied that candidates would study simultaneously for their graduate research and for their BA finals, and this eventually led to criticism of the program. In 1886 a statement by Dalhousie University called for elimination of "the provision that the student has to become a candidate for a degree because of loss of time."[4] Nine years later a report prepared by former holders of the scholarship suggested that the strain of preparation had undermined the health of some candidates.[5] In 1897 the program was abandoned. Nevertheless, it had provided support for eighteen students, of whom seven would go on to be professors in Canadian universities. Among the seven of the eighteen who specialized in science, three would pursue their careers in Canada: S.W. Hunton taught mathematics at Mount Allison; W.L. Goodwin, chemistry at Queen's; and J.G. MacGregor, physics at Dalhousie.[6]

Because it was awarded to only one student per year, the Gilchrist Scholarship could not have any major overall effect on the development of post-graduate studies in Canada, except perhaps indirectly, as illustrated by the career of MacGregor, who passed on to his post-graduate students the benefit of the training in research that he had received in Edinburgh. Oddly enough, the opening of Johns Hopkins University in 1876 would have more effect on the development of graduate study in Canada than did the scholarship offered by the "mother country."

DOCTORAL STUDIES AT TORONTO AND MCGILL

In order to widen the clientele of their new, research-based doctoral programs, American universities offered scholarships to their most able students regardless of nationality. Johns Hopkins began this trend, followed by Cornell, Harvard, and Chicago. Canadian students, just as aware as their American counterparts of the advantages to be derived from studying in these programs, did not hesitate to cross the border in large numbers. During the last quarter of the nineteenth century the four universities mentioned had an enrolment of almost three hundred Canadians, and more than one-third of these received financial support.[7] Of the total number close to one-third came from Toronto, and these students accounted for half of the scholarships received. The University of Toronto, especially hard hit by this exodus, was not surprisingly the first to react.[8]

In 1883 the University of Toronto offered nine post-graduate scholarships of five hundred dollars each, equal in value to those available at Johns Hopkins.[9] However, whereas at Johns Hopkins the recipients devoted all their time to the preparation of a doctoral thesis, at Toronto they had to assist their professors in teaching duties. Because the university was in a precarious financial situation, departments rapidly came to use these funds simply to hire instructors and demonstrators. At US universities the award of scholarships was tied to a well-defined course of study leading to the doctoral degree. At Toronto, by contrast, the awards represented a hasty effort to ward off the dangers of competition from the south, and there was no genuine structure of post-graduate instruction. Because the work of the scholarship-holders did not lead toward a doctoral diploma, the net result was to intensify the trend toward study in the United States. There, the same work would result in acquisition of the PHD degree, increasingly necessary for anyone aspiring to a university career.

A first step toward a real solution to student emigration could only be the creation of a doctoral program. In Toronto, the most ardent proponent was James Loudon, who had advocated the development of scientific research since the 1870s. In his presidential address before members of Toronto's Canadian Institute in 1877, Loudon had already vaunted the merits of the German model of research and dreamed of the day when Canadian universities would become a home for a large number of researchers, who would not be forced to teach many courses at once and would be able to devote themselves to teaching and research in their specialty.[10]

Loudon remained for a long time in a minority position, as the president of the university, Daniel Wilson, was opposed to the idea of a doctoral program. When he became university president in 1892, Loudon was finally in a position to realize his project. Supported by the youngest professors, in particular the biochemist A.B. Macallum, a University of Toronto graduate who had obtained his PHD at Johns Hopkins, Loudon succeeded in establishing a doctoral program in 1897.

Administered by a senate committee, the program was directed by Macallum, who did not leave the position until twenty years later in order to become first chairman of the NRC and, in this capacity, continue to encourage the development of scientific research in Canada.[11] In July 1897, in the first issue of the *University of Toronto Monthly*, President James Loudon stated clearly the argument that he had been pressing within the university community for twenty years: "The old ideal of a University as merely an institution for the transmission of knowledge is passing away. This ideal is that of the College as contrasted with the University proper which has the additional function of adding to the sum total of knowledge by original research."[12] In the same year the university calendar announced that "the degree of Doctor of Philosophy has been established for the purpose of encouraging research in the University." Correspondingly, the University of Toronto Series was launched as a means of publishing theses and other works arising from the research of professors.

This, Canada's first doctoral program, initially had no formal structure of courses. It evolved under American influence to take its final form during the 1910s. In 1904 the master's program was modified to include presentation of a paper embodying the results of original research.[13] According to A.B. Macallum, these changes were long overdue; if implemented fifteen years earlier, he believed, they would have given direction to the work of earlier scholarship-holders, to the great benefit of the development of graduate studies at the university. Macallum was well placed to make this judgment, since he had proceeded after graduation from the University of Toronto to obtain a PHD at Johns Hopkins in 1888.[14]

At McGill the first modifications to the master's program were made in 1899, and the PHD degree was not adopted until 1906.[15] Even this institution, which had always enjoyed a privileged relationship with the major British universities and had recruited most of its professoriate in Britain, had no alternative but to adapt to North American trends. McGill's doctoral program, like that of the University of Toronto, was heavily influenced by the US model and led in 1922 to creation of a Faculty of Graduate Studies.[16] In the same year the University of Toronto adopted a similar structure under the title of School of Graduate Studies.[17] In 1926, the two institutions joined the Association of American Universities – founded in 1900 to co-ordinate the post-graduate offerings of US universities – a logical culmination of their increasing assimilation into a North American pattern.[18]

The convergence between Canadian and American universities was judged inevitable by Loudon, who in 1902 wrote that "such is the geographical position of Canada with regards to the United States, and such the community of social and intellectual life, that the universities of these two countries must inevitably develop along parallel lines."[19] This speech, given as president of the Royal Society of Canada, was in some respects Loudon's intellectual testimony. Four years from retirement, Loudon thus returned to the theme he had broached before the members of the Canadian Institute twenty-five years earlier. A tireless promoter of university research, Loudon gave a final eulogy to the German model. Criticizing British universities for their indifference to research, he reminded his audience that despite the progress in Canada in the previous ten years, which he attributed to American and German influences, "organized research in Canadian universities, as a definite system, can scarcely be said to exist as yet."[20]

Loudon's dream would not in fact become reality until after the First World War. It fell to his young colleague and protégé, Macallum, to contribute to the establishment of a university research system in Canada by becoming chairman of the NRC in 1917 and by creating shortly thereafter a system of scholarships for master's and doctoral students. Until then, McGill awarded only one doctorate a year for all disciplines;[21] at Toronto, this mean was valid for the period 1896–1907 and doubled in the following decade.[22]

THE ADVANTAGES OF COLONIALISM:
THE 1851 EXHIBITION SCHOLARSHIP

Before establishment of the NRC, students interested in a scientific career could expect no significant financial aid from Canadian universities. Fortunately for them, England created scholarships designed specifically for young scientific graduates aspiring to training in research. In the mid-nineteenth century a movement in favour of the development of industrial research emerged in England. First taking shape in the report of the 1850 commission of inquiry on Oxford and Cambridge universities, the movement gathered force with the appointment in 1870 of the Devonshire Commission on technical education, and resulted in 1890 in the creation of scholarships to encourage the training of scientists who would assist the industrial development of the British Empire.[23]

In 1890 the Royal Commission for the Exhibition of 1851 announced establishment of new scholarships. Charged with the management of the accumulated profits of the Great Exhibition, the commission had already assisted such national institutions as the South Kensington Museum and the Royal College of Arts and Science. Now a study committee was quickly established. Following wide consultation, its chairman, John Playfair, recommended launching of a scholarship program similar to that already developed by Jean-Baptiste Dumas at the École pratique des hautes études in Paris. The scholarships would amount to £150 a year and would be open to British subjects under thirty years old who had demonstrated during their university studies a special aptitude for, and interest in, research in pure or applied science. Applicants were free to pursue their studies for two or three years in Britain or elsewhere in the world.[24] Of the twenty scholarships to be offered each year from 1891, six were to be awarded to parts of the empire outside Britain, and two of these were reserved for Canada.

The scholarships were further designed exclusively for the scientific disciplines: biology, chemistry, engineering, geology, and physics. From 1891 to 1917 they played an important role in the training of Canadian scientists. R.T. Glazebrook, director of the program, summed up in 1930 the significance of the 1851 Exhibition Scholarships, in the preface to a report that analysed the career patterns of those who had received the awards:

Established at a time when the field was still untouched by any system that carried training beyond the limits of ordinary degree curricula, these scholarships have undoubtedly given a great and much needed impetus to postgraduate study. They certainly played an important part in raising the standard of teaching in the younger Universities and Colleges of the Empire, and the hope, originally entertained, that in the course of time, there would arise, in the yearly allocation within the Empire of some eighteen scholarships, a body of well-trained men of science who would be able to extend the bounds of natural knowledge, has since been abundantly realized.[25]

This judgment is borne out by the case of Canada and by the evidence from the four universities that participated in the program: McGill and the University of Toronto shared one scholarship, each awarding it in alternate years, as did Queen's and Dalhousie. Although McGill and Toronto were already relatively well equipped in the sciences, Queen's and Dalhousie (integrated into the program in 1893 and 1894, respectively) were much less so. Especially at Dalhousie, science teaching benefited greatly from involvement with the scholarships.

A committee appointed by Dalhousie's board of governors to study the implications of the new initiative reported that scientists at the university "have never been authorized by the Board of Governors to make a greater annual expenditure on ... [physics and chemistry laboratories] than is necessary for conducting the ordinary university classes ... and that while at present some facilities for research in a few very narrow departments can be afforded ... it will in two or three years be impossible not only to provide these meager facilities but even to provide practical instruction of any kind."[26] The committee estimated that an annual investment of $100 in each laboratory would be enough to halt the deterioration, but that any major improvement would be produced only with new annual expenditures of $300 to $400. It concluded that the offer of an 1851 Exhibition Scholarship could be accepted by Dalhousie for the years 1894 and 1896 if $100 per year were spent on each laboratory but that "the periodical repetition of the Commissioners' offer cannot be expected unless an additional annual expenditure of about $100 or $150 on each laboratory can be provided for."[27]

As a result, the governors authorized the university senate, from 1894, to disburse up to $400 annually for laboratory improvements. The new expenditures allowed the physicist J.G. MacGregor, for example, to buy new apparatus that he had been denied for some years and thus to "afford greater facilities for original research."[28] Because of these new investments, MacGregor was able to initiate his students to research, as we saw in chapter 1. When he left Dalhousie in 1901 to succeed his mentor P.G. Tait at Edinburgh, MacGregor had trained at least eight students. All had profited from the 1851 Exhibition Scholarship, and the majority of them earned a doctorate in physics or chemistry from an American university. They subsequently found employment in Canada or the United States.

By contrast with Dalhousie, McGill and the University of Toronto were able to meet the requirements of the scholarship without difficulties. McGill nominated its first 1851 Exhibition Scholar in 1891, and Toronto did the same the following year. From 1893 onward McGill and Queen's named scholars in the odd-numbered years, while Toronto and Dalhousie did so in the even-numbered years. From 1891 to 1914, thanks to this financial aid from Britain, forty-seven Canadian students were able to acquire scientific training in the leading research laboratories of the world.

Analysis of these recipients (Table 2.1) shows that by no means all of them were attracted to study in Britain. Scholars from Toronto and

Table 2.1
Destinations of 1851 Exhibition Scholars by Institution of Origin,
All Disciplines Included, 1891–1914

University of Origin	United States	Great Britain	Germany	Other Countries	Total*
Dalhousie	10	1	1	0	12
McGill	4	7	3	0	14
Queen's	6	2	4	0	12
Toronto	1	4	7	0	12
Total	21	14	15	0	50
% Canadian Recipients	42	28	30	0	100
% Colonial Recipients†	20	52	20	8	100

Sources: Record of the Science Research Scholarship of the Exhibition of 1851 (London 1930); R.M. MacLeod and E.K. Andrews, "Scientific Careers of 1851 Exhibition Scholars," Nature 218 (15 June 1968), 1013–14.
* The total is larger than the number of recipients (47) because some visited two countries with the same scholarship.
† The "colonies" included were Australia (before and after creation of the Commonwealth in 1901), Canada, New Zealand, and South Africa (before and after union in 1910).

McGill frequently did go there, but those from Queen's and Dalhousie tended to go to the United States. Overall, the Canadians were drawn much more to the United States and Germany than were recipients from other parts of the empire, who overwhelmingly opted for Britain. There was also variation according to discipline. A large majority of students in chemistry went to Germany and used the scholarship to study at the famous laboratory of Wilhelm Oswald in Leipzig. Physicists from McGill and Toronto normally went to Britain, but those from Dalhousie most often studied at the major American universities.

These two different directions reflected the histories of the various departments and the varying networks or relationships that they had built over the years. The evidence also indicates that, contrary to conventional interpretations, the colonial relationship between Britain and Canada did not prompt Canadian science students to gravitate inevitably toward Cambridge. A significant factor affecting these choices was the shortage of doctoral programs: those students who studied in the laboratories of British universities would attain only a BA or MA degree, while those who chose US or German universities would return with a PHD. This discrepancy was often noted with disapproval by Canadian university presidents.[29]

Table 2.2

Area of Specialization of 1851 Exhibition Scholars by Institution of Origin, 1891–1914

University of Origin	Physics	Chemistry	Biology	Geology	Engineering
Dalhousie	4	7	0	0	0
McGill	6	4	1	0	2
Queen's	2	2	1	6	0
Toronto	6	5	1	0	0
Total	18	18	3	6	2
% Canadian Recipients	38	38	6	13	4
% Colonial Recipients	49	24	10	9	8
% British Recipients	31	54	7	3	5

Sources: As for Table 2.1.

As regards the choice of disciplines, Canadians conformed to the general pattern. Physics and chemistry (Table 2.2) were far ahead of the rest. The British, perhaps responding to industrial needs, awarded almost twice as many scholarships to chemists as to physicists. Among the imperial recipients that proportion was reversed. Within Canada the universities of Toronto and McGill – with the large endowments – had well-established departments in several of the disciplines, and this was reflected in the choices of their candidates. At Queen's the majority opted for geology; since the opening of the School of Mining in 1893 this had been the university's chief area of scientific specialization.

Although only half the scholarship recipients subsequently carried on their scientific careers in Canada while the others found employment in Britain or the United States, it is not justifiable to conclude, with Robin Harris, that the 1851 Exhibition Scholarship either did not advance or may have retarded the development of graduate studies in Canada.[30] In physics, for example, eleven of the eighteen award-holders returned to Canada, and ten continued to be active in research, nine in the universities and one at the meteorological office of the dominion government. Seven became fellows of the Royal Society of Canada and can be regarded as having played an active role in the development of physics in the country. To be sure, the limited number of scholarships available – two each year – and the fact that they were used for study outside Canada combined to ensure that the program could never supply comprehensive, long-term stimulation to scientific research at Canadian universities. Nevertheless, it is fair to conclude that, between 1891 and 1917, the 1851 Exhibition Scholarships helped to form the first nucleus of Canadian scientific researchers and that these early scientists were then instrumental in generating research activity at Canadian universities.

GROWTH OF THE CANADIAN
UNIVERSITY SYSTEM

While there were few university positions available before 1900, the opening of universities in the Canadian west and the rapid growth of their student population soon created more jobs. In physics between 1900 and 1914, the number of positions grew more quickly than the Canadian production of doctorates, and universities were forced to recruit professors who had earned their diplomas abroad, notably in the United States. This deficit, which continued until the end of the Second World War, allowed repatriation of a good number of researchers trained in American universities.

As we can see in Figure A1 (in the Statistical Appendix to this volume), the number of positions for physics professors (from assistant to full professor) rose from six in 1900 to twenty-one in 1914. Smaller institutions, such as Acadia, Mount Allison, and St-Francis-Xavier, played only a marginal role in the development of physics in Canada. They limited their activities to teaching and served mainly as waiting rooms for those who, holding a doctorate, were waiting for positions in a better-equipped university – the only way to improve their position in the scientific field (by having access to the means of production necessary for publication of scientific papers).[31] The rate of turnover of professors was higher in these institutions, none of which (not even the University of New Brunswick) belonged to the 1851 Exhibition Scholarship program.

Given that almost half of the increase in positions between 1900 and 1914 took place in western universities, we shall briefly recall how these universities developed.

Although founded in 1877, the University of Manitoba did not have its own professors and students until 1904.[32] As the provincial government refused to increase the university grant, the governors were forced to turn to private sources in order to hire the first science professors (botany, chemistry, mathematics, physics and physiology). They approached, among others, Lord Strathcona, the financial and railway baron, who agreed to donate $5,000 a year for four years in order to help defray the salaries of these new professors, fixed at \$2,500 each.[33] Thus Frank Allen, a graduate of the University of New Brunswick (BA 1895; MA, 1897), who had been awarded a PHD by Cornell in 1902, was hired to teach physics. In 1909, the student population having almost doubled, the university called on the services of another physicist, Robert Kenneth McClung, one of Rutherford's first collaborators at McGill, where he had obtained his DSC in 1906. Between 1906 and 1909, he was one of those "waiting" at Mount Allison.

In 1908, shortly after the creation of the province, the University of Alberta opened its doors. Four years later, while assistant professor at McGill, Robert William Boyle was called on by his colleague H.M. Tory to become professor of physics. A McGill graduate, like Tory, Boyle had been the first to be awarded a PHD by that institution.[34]

Rivalling its twin province, Saskatchewan opened its university in 1909, and, one year later J.L. Hogg, apparently dissatisfied with his position at McMaster University (which was then located in Toronto), left to teach physics in Saskatchewan.[35] He had received his undergraduate degree from the University of Toronto in 1899 and a PHD from Harvard in 1904. He had taught three years at McMaster and was immediately replaced by Henry Franklin Dawes, another Ontarian, who obtained his PHD from Toronto in 1918. In 1914, Hogg was joined at the University of Saskatchewan by the American Albert Edward Hennings, a recent University of Chicago graduate (PHD, 1914). Hogg left the department in 1919, after having been involved in an intrigue which had sought the resignation of the president, Walter C. Murray. The latter had won the dispute, and three department heads were forced to resign.[36] Hogg subsequently disappeared from the Canadian scientific community. Hennings left in 1918 and resurfaced at the University of British Columbia the following year.

The last major university to see the light of day was the University of British Columbia, in 1914. It had little difficulty in recruiting professors, as in 1906 the governors of McGill had put H.M. Tory in charge of opening a branch of that university in the province.[37] Thus James Grant Davidson (BA, Toronto, 1900; PHD, California, 1907), who was teaching at what was then known as McGill College, became professor at the new university.[38]

The student bodies of these four western universities grew rapidly, and, at the end of 1928, each had at least three physics professors. This quantitative change was accompanied by qualitative improvements, which in 1921 were recognized by the commissioners of the 1851 Exhibition Scholarship, who, profiting from reform of the rules, added these four universities to the list of those invited to recommend candidates for the bursary.

Meanwhile, the physics department at Queen's had developed considerably, particularly since the arrival of the American Arthur Lewis Clark (PHD, Clark, 1904), who came to join William C. Baker in 1906.[39] In 1928, the department had five professors – the same number as Toronto and three less than McGill.

Let us now take a closer look at this population of physicists working in Canadian universities. The forty-seven positions available during the period 1900–28 were occupied by only sixty-six candidates, which indicates very little migration by occupants. In fact, of the twenty-one physicists who had a position in 1914, only seven had left by 1921. Similarly, of the thirty-six physicists employed in 1921, only four had left by 1928. This stability was clearly related to the fact that two-thirds of this population was of Canadian origin.[40] There were only seven Americans, seven Britons, and one Australian. Among those who were present in 1900, Cox and Rutherford had returned to England, in 1909 and 1907, respectively.

While British and American physicists usually received their training in their own countries, the situation was more complex for Canadians. An analysis of the origin of PHDs (or DSCs) held by Canadian professors shows that an equal number obtained their PHD in Canada and the United States, while few had earned a doctorate in Britain.

The cost of studying abroad helps explain why Britain attracted so few Canadian students. However, the decisive factor was the structure of British universities, which did not recognize colonial diplomas and were poorly adapted to foreign students' needs. We have already seen that this system had been criticized in the 1880s by holders of the Gilchrist Scholarship. During the 1900s and 1910s, the principal of McGill and the president of Toronto returned continuously to this question by emphasizing that if British universities did not adapt to the American system, Canadians would emigrate to the United States and would be a loss to the empire. Nonetheless, it was necessary to wait until 1919 for Cambridge and Birmingham to establish PHD programs.[41]

A finer look at the professors' trajectories shows that, according to the origin of first university diploma, Canadians followed different routes in the pursuit of their careers. As the analysis of the 1851 Exhibition Scholars has shown, Dalhousie graduates had a strong tendency to continue their studies in the United States, which fact can be explained easily by the absence at Dalhousie of a strong PHD program but especially by the existence of privileged relations with American physicists. Thus MacGregor had contact with E.L. Nichols at Cornell and sent his first two students there, in 1894 and 1896. Another of his students, Arthur Stanley Mackenzie, who taught at Dalhousie beginning in 1905, had studied at Johns Hopkins (PHD, 1894) and had belonged to the founding committee of the American Physical Society in 1899.[42] Finally, the American H.L. Bronson, who came to Dalhousie in 1910, had studied at Yale.

The trajectory of McGill students was different. As McGill had awarded the DSC since 1898 and the PHD since 1906, graduates tended to undertake more or less extended visits to Cambridge or Manchester before returning to McGill to receive their DSC or PHD. Robert K. McClung, for example, used his 1851 Exhibition Scholarship for research at Cambridge between 1901 and 1904 and received a DSC from McGill in 1906. In contrast, Robert W. Boyle received his PHD from McGill in 1909 and subsequently went to Manchester to work with Rutherford, thanks to the same bursary. At the University of Toronto, the situation was analogous. Elie F. Burton spent 1904, 1905, and 1906 at Cambridge and at Birmingham and, returning to Toronto, wrote a thesis on the physical properties of colloidal solutions for which he received his doctorate in 1910.

Of the almost fifty physicists of Canadian origin who held a position between 1900 and 1928, a dozen studied at Cambridge or Manchester. Six of these benefited from the 1851 Exhibition Scholarship. This bursary naturally inclined Canadians toward Britain. However, the Cambridge school dominated the field of physics. Moreover, the fact that many Dalhousie graduates used this same bursary to go to the United States indicates that the trajectories followed by Canadian physicists were not mechanically determined by colonial relations. Contrary to Australian physicists, who always turned toward the mother country,[43] Canadians took advantage of the proximity of the large American universities. This closeness was more important than political ties with Britain, and Canadian physicists with American PHDs have always been more numerous than those with British PHDs.

THE SEARCH FOR SPACE, TIME, AND GRANTS

In Canada, prior to the First World War, the number of university professors in physics had grown to about twenty, but most of those who could find time to work on research projects had few resources and were forced to use apparatus acquired for laboratory teaching. At Dalhousie, for example, J.G. MacGregor, who had trained at Edinburgh under Tait in the mid-1870s and specialized in the physical properties of aqueous solutions, worked for twenty years without a real research budget. For some time he had to use his summer months to travel to Edinburgh, where, "through the kindness of professor Tait," he could make use of the rich stores of the Natural Philosophy Laboratory of the

University of Edinburgh to further his research.[44]

MacGregor finally received a grant of $100 from the Royal Society of Canada in 1900. Two years earlier the society had saved part of its annual dominion grant of $5,000 and applied the surplus, "in view of stimulating scientific research," to help research projects undertaken by members of the scientific sections.[45] In 1899 the money was used by section IV, biology and geology; next year it came to section III, and MacGregor, a charter fellow of the society, received the grant, under the condition that "the results of such researches must be reported to the Section with the object of having them published exclusively in the Transactions."[46] It was, however, a little late to help MacGregor, who left Dalhousie in 1901 to succeed Tait at Edinburgh. Unfortunately, there was no spare money in the following years, and no more grants were made available to promote research.

In contrast to the lack of funds at Dalhousie, the McGill physics department was lavishly supported from 1893 to 1907, thanks to the generosity of Sir William C. Macdonald, the tobacco magnate. Opened in 1893, the Macdonald Physics Building was one of the best equipped in the world. In his report to the university for that year, director John Cox mentioned that "the unusually complete set of electrical standards and instruments for comparison and the new instruments for thermometry and pyrometry, indicate that the laboratory may do useful work in these two branches of physics at no distant date."[47] Cox, though not a researcher himself, was anticipating that the contributions of his newly appointed colleague, H.L. Callendar, would add to the science of thermometry.[48] In 1892 Macdonald contributed $40,000 for the maintenance of the laboratory and the salary of a janitor and a technician. Four years later he added $110,000 to the fund to provide new instruments and an instrument maker.[49]

When new problems arose, Macdonald was nearly always there to solve them. In 1900 the growth of the student population required the hiring of new demonstrators for the physics laboratories, but the university's resources were not sufficient to meet this need. Macdonald accordingly gave $2,000 to pay the salary of two new assistants.[50] This contribution also helped research, for most demonstrators were associated with the research projects of the professors.

In addition to salaries, money was often required for instruments. In 1902, for instance, Ernest Rutherford and Frederick Soddy were working on the nature of the radium emanation (now known as radon) and needed liquid air. "On learning that several investigations were at a stand still for want of a supply of liquid air, Cox reported, Sir W.C. Macdonald further caused a complete liquid air plant on Dr Hampson's plan to be purchased for the Physics Building at the cost of $1,250." The very night the apparatus was installed, the gaseous nature of the emanation was finally proved beyond doubt.[51] The same year, Macdonald provided $500 to pay for 80 milligrams of radium and added "a special contribution towards the expense of the large number of papers published in the various scientific periodicals."[52] In 1903, Rutherford was working on the deflection of ions in a magnetic field. Once again, Macdonald provided $2,000 for the year "for the purpose of experimental work and new apparatus," and Rutherford was able to buy a large electromagnet.[53]

Notwithstanding the great success of Rutherford's research program at McGill, the physics department was created primarily to teach physics to engineering and medical students, and the growing costs associated with the expanding student population left little money for research. It seems clear that without Macdonald's patronage, Rutherford's research activities would have been much reduced.

In the years that followed, the situation became more and more difficult. In 1906 Cox noted that he had no money to hire demonstrators but mentioned with satisfaction that "research has not, however, been hampered thereby, for the special research fund contributed by Sir William Macdonald some four years ago has been carefully husbanded, so that there is still a balance of $400."[54] When Rutherford left McGill in May 1907, the epoch of patronage that helped him to stay at the forefront of atomic physics was gone.

Succeeding Rutherford as Macdonald Professor of Physics was Howard Turner Barnes, a former student of Callendar's, whose work on the properties of ice was of a more applied character. He received several grants from the dominion Department of Fisheries and Oceans, which paid for instruments and, sometimes, an assistant.

Other succeeded in obtaining research funds from abroad, in particular, the United States. At Queen's University, Arthur Lewis Clark, an American, received a grant from the Rumford Fund of the American Academy of Arts and Sciences, permitting him to do research on the thermodynamic properties of solutions;[55] the grant enabled him to hire an assistant and to purchase some equipment. Several years later, Louis V. King, who had become assistant professor at McGill, also received funds from this organization as well as from the Royal Society of Canada.[56]

TEACHING V. RESEARCH

The rapid growth in the number of students in the universities during the century's first decade – which was not followed by a corresponding increase in budget – reduced the already meagre economic resources allocated for laboratory equipment (which indirectly helped in the development of research) but also resulted in a struggle for space which harmed research.

At McGill in 1909, the thermodynamics laboratory was forced into modifications allocating "the larger proportion for the medical students and reserving one for research."[57] The following year, Barnes, who had directed the laboratory since Cox's retirement in 1909, complained of a lack of space and noted that "it is inevitable that the original work done in the building must suffer in consequence."[58]

This kind of conflict between teaching and research can also be seen at other Canadian universities. At the University of Toronto in 1908, Elie F. Burton, then assistant demonstrator in physics, wrote to the head of the department, J.C. McLennan, to complain that the rooms he was using for his research projects had been requisitioned as classrooms: "My own research is now in progress in one of these rooms, and if this room is taken, and my apparatus dismantled, I must abandon my research." He reminded McLennan: "When I accepted my position here, instead of going to Princeton University, you will remember that it was on the distinct understanding that I should have reasonable time and advantage for my research work."[59] Recalling the crowded situation that had prevailed prior to the opening of the new building in 1907 and the inconveniences occasioned by construction, he concluded: "Now I feel that if we are to continue such hand-to-mouth existence, a University position in Toronto has very little advantage over that of an ordinary High School Teacher."

Someone like McLennan could understand such complaints. Twelve years earlier, as a demonstrator, he had made a similar demand to James Loudon. At the beginning of the 1897 university year, the young McLennan had pointed out that the growth in student numbers was such that "it has become quite impossible to carry out my laboratory work with desirable efficiency unless more room is placed at my disposal." He added: "I desire also to undertake from time to time special investigations of my own in some room in which my apparatus shall be free from disturbance by the students."[60]

In fact, Burton's letter had apparently been solicited by McLennan, who had hoped to use it in approaching the president of the university. In his letter to Robert A. Falconer, dated the same day as Burton's letter, McLennan went to the heart of the matter: "I may state that after giving the whole question the fullest consideration, my opinion coincides with Mr. Burton's on the necessity for leaving undisturbed the rooms ... at present in use by Mr. Burton and his students."[61]

At Queen's University, the demand for space and time to do research did not emerge until after the hiring of Arthur L. Clark in 1906. Before that, the position had been occupied by D.H. Marshall, who, since 1882, had devoted all his time to teaching. At the turn of the century he had been assisted by Norman Ross Carmichael, a Queen's graduate. The latter had taken advantage of the 1851 Exhibition Scholarship to study electrical engineering at Johns Hopkins between 1893 and 1895. He died prematurely in 1908. At this time, the financial situation at Queen's was precarious, and the dean of the Faculty of Applied Sciences complained of the lack of resources in the laboratories and deplored the fact that the university "had no William Macdonald to come to its aid."[62]

When he arrived to direct the department, Clark had just finished at Clark University, where he had completed a doctorate under A.G. Webster. Very early on, Clark insisted on the importance of research, and in his report for the academic year 1911–12 he wrote: "As was pointed out last spring, original investigation should be more adequately provided for. I might mention however that considerable research is going on at present."[63] He was referring to his own work on the thermodynamic properties of certain substances and to William C. Baker's work on radioactivity. Having been named assistant professor in 1908, the latter had been initiated into research during a stay at the Cavendish Laboratory between 1900 and 1902 paid for by the 1851 Exhibition Scholarship.[64]

Clark noted in 1912: "Some of the very distressing features of our attempt to carry out research are the lack of private rooms and the scarcity of apparatus so much of which is needed for the regular class work."[65] Aware that research could not yet be ranked ahead of teaching, he added: "Our first duty is to our students and if apparatus is needed for the undergraduates' work it must be used even if the research be sacrificed."[66] Clark pointed out: "It is to be hoped that research may be more adequately provided for, so that Queen's University may take her place amongst Canadian Institutions in the way of original scientific investigations as she has in other lines of University activity."

This appeal to institutional pride and prestige, which is the expression of a form of competition among universities, is also one of the strategies employed by researchers to convince their institution of the utility, symbolic at least, of their work. At McGill, for example, John Cox continually drew the attention of university authorities to the fact that Rutherford's work increased McGill's reputation around the world.[67] In the same way, at Toronto, in the mid-1920s, the discovery of insulin created enormous publicity for the university, and subsequently nobody could ignore research, which had become an important element in the competition between universities. Falconer took advantage of the discovery of insulin to write to the premier: "The discoveries made in our laboratory have carried the name of Toronto far and wide."[68] If at the

beginning of the century, McGill had its Rutherford, Toronto found its
Banting. When individuals leave (such as Rutherford leaving McGill)
or die (such as Banting), institutions can continue to hold on to the
famous names through eponymy: the Banting Institute and the
Rutherford Laboratory.

The end of the First World War led to an increase in students, which
necessitated the hiring of new professors. In this context, the
consolidation of the group of researchers, until then few in number,
could come only through the imposition of new hiring criteria. In 1919,
Clark wrote: "It is to be hoped that in making appointments in the future
a fair proportion of the men selected will be of the research type."[69]
The departmental structure favoured expansion of the research faction
within the professorial corps, since, more and more, presidents of
universities endorsed department directors' choices. After the war, it
became rather rare for a professor to be hired if he did not have a
doctorate. At Toronto, for example, one can see this criterion of choice
at work earlier than at Queen's, for the category of researchers had
appeared at least a decade earlier. Expressing his opinion about a
candidate for a position in biochemistry, A.B. Macallum wrote: "He is a
splendid teacher, but before all else a researcher."[70]

THE FIRST WORLD WAR AND
THE GROWTH OF RESEARCH

"In 1906," remarked H.J. Cody to the Royal Society of Canada forty years
later, "research did not occupy its present position in the thought and
practice of our Canadian Universities."[71] For a variety of reasons,
however, that situation was about to change. Part of the explanation, as
already shown, lay in the increasing presence in Canadian universities
of professors trained in research who intended to continue as active
scientists.

Even more important was the influence of the First World War. The
wartime conjuncture gave scientific research national importance:
science could no longer be the sole responsibility of a handful of
scientists at a few universities. During the 1910s the movement for
industrial research also gathered strength in Canada, prompted by the
industrial establishment, working through the Canadian Manufacturers'
Association (CMA) and the Royal Canadian Institute, which acted as a
bridge between industry and the universities. The war made it clear how
completely Canadian industry had depended on equipment and
technologies imported from Europe.[72]

The pressure exerted by the industrial milieu, supported by the
presidents of the major Canadian universities, led to a first meeting with

the minister of industry and commerce, Sir George Foster, on 25 May 1915.[73] Among the dozen individuals invited by the minister to discuss measures to develop industrial research in Canada were the presidents of McGill, Toronto, and Queen's, the physicists H.T. Barnes of McGill and J.C. McLennan of Toronto; and the chemists R.F. Ruttan of McGill and W.L. Goodwin of Queen's. Discussion turned to creation of a commission to study industrial research, but no decision was taken. Occupied by more urgent problems related to the war effort, Foster continued to reply that he would act as soon as possible.

In the mean time, in July 1915, England had established its own research council to co-ordinate industrial research.[74] As Canada was part of the empire, the directors of the council naturally turned to Canadian universities for co-operation in the war effort. This request in turn provoked William Peterson, principal of McGill, to write to Foster, in May 1916. Unless concrete proposals were made, he intended to suggest to other universities, during the meeting on 23 and 24 May of the National Conference of Canadian Universities at McGill, that they associate with the British Research Council.[75] The fear of Canadian universities dealing directly with Britain without going through the Canadian government was a stimulus to action. One month later, an order-in-council created a scientific and industrial research subcommittee of the Privy Council. In November, this committee created a consultative council, based on the British model, under the name of the Honorary Advisory Council on Scientific and Industrial Research.

Composed of eleven members and a secretary all chosen by the dominion government, the Research Council was controlled by academics. The biochemist A.B. Macallum, who had helped develop scientific research at the University of Toronto, was named administrative chairman. Among the other members were F.D. Adams, dean of the Faculty of Applied Sciences of McGill; his colleague R.F. Ruttan; J.C. McLennan from Toronto; and A.S. Mackenzie from Dalhousie. A physicist, Mackenzie had been a student of J.G. MacGregor's and had succeeded his mentor in the Monro Chair of Physics in 1905. A Johns Hopkins graduate, he became president of his university in 1919. Another university president, Walter C. Murray, from Saskatchewan, was a member of the council, as was S.F. Kirkpatrick, a professor of metallurgy at Queen's – making, in all, a total of seven academics.[76] The other members were R.A. Ross and Arthur Surveyer, Montreal consulting engineers, and T. Bienvenu, vice-president of the Provincial Bank.[77] The superintendent of the Dominion Water and Power Branch, J.B. Challie, acted as secretary.

The predominance of the university milieu was difficult to avoid. As industrial research was practically nonexistent in Canada, the academics, acting through the Royal Canadian Institute (presided over by

J.C. McLennan in 1916) and supported by the CMA, had pressed the government to act, convinced that industrial research could be based only on the development of university research. In fact, during the first third of the twentieth century, the principal organizations involved in the development of scientific and industrial research in Canada brought together a relatively small number of individuals to make decisions, either at the Royal Society of Canada, the National Research Council (the renamed Honorary Advisory Council), or the National Conference of Canadian Universities. Moreover, their assiduity at council meetings meant that in practice they directed the NRC.

Named on 29 November 1916 by an order-in-council, the members of the Honorary Advisory Council held their first meeting in Ottawa from 4 to 6 December. Since the main function of the council was to co-ordinate research activities, both university and industrial, its first task was to take stock of the existing situation. In its first annual report, tabled in April 1918, the council presented a study, "The Status of Research in Canada," wherein it exposed the underdeveloped state of scientific and industrial research. The report insisted that "unless vigorous action is taken, and soon, to reorganize our industries on scientific lines, Canada will face a very serious industrial crisis in the years following the war."[78] The authors pointed out that the Canadian situation was comparable to that in other countries in the empire and that even Britain had neglected to encourage scientific and industrial research until mid-1915.[79]

After having recalled the evolution of attitudes toward scientific research in different countries, the report noted: "Everywhere throughout the Empire there has occurred in regard to the attitude of the State towards scientific research, a complete break with the past."[80] It mentioned that in the eighteen Canadian universities the science faculties were, for the most part, of recent origin and that most were badly equipped and poorly organized. According to the authors, the annual budget of the Massachusetts Institute of Technology exceeded the annual expenditures of all the Canadian science faculties combined. With regard to industrial research, the report estimated that thirty-seven companies had research laboratories which represented a total investment of $237,000. Moreover, if companies had sought to hire researchers, they would have had difficulties finding Canadian manpower. The council reported that less than a dozen doctorates in pure science had been awarded by Canadian universities and estimated that there were less than fifty Canadian researchers.[81]

For university members of the council, these facts were not unknown and only served to justify their action, which had begun long before the report had been written. At the first meeting, it had been decided that $12,000 would be made available for a bursary program in higher education; during the third meeting, a committee had been formed to elaborate an aid program for scientific and industrial research, two initiatives to which we shall shortly return.

Several months before being named director of the NRC, Macallum, as president of a committee on graduate education of the National Conference of Canadian Universities (NCCU) meeting at McGill in May, had asserted: "The two great needs of Canadian Graduate Schools were scholarships and increased library facilities, because it was through these that the American universities were able to attract so many of our Canadian graduates."[82] Named chairman of the NRC and surrounded by academics who could not help but share this point of view, Macallum was now in an excellent position to realize immediately the first of these objectives, at least for the scientific disciplines. As for the development of industrial research – the NRC's primary objective – it could, according to Macallum, be based only on the existence of faculties of graduate studies, "thoroughly equipped like an American Graduate School of first rank," within several Canadian universities.[83] For the first director of the NRC, the primary task for the universities was to prepare students for an industrial career by basing research on problems of both pure and applied science and not limiting their training to the restricted problems of industrial research. In summary, "on the general principle of utility, as well as because of the ideals the student who is training for industrial research should [for the period of the fellowship] concern himself with problems in pure and applied science."[84]

The NCCU's committee, created in 1919 "to consider in what way the universities may best cooperate in the development of scientific and industrial research," also accepted this point of view.[85] Presided over by the physicist A.S. Mackenzie and including the physicists A.L. Clark and A.S. Eve and the chemist Douglas McIntosh (a student of J.G. MacGregor's), this committee wrote in its report:

In the co-operation of all possible agencies for the organization of the prosecution of scientific research for the industries of the country, the prime part which the scientific departments of our universities should take is ... the development of the spirit of research in them [the students and the public] and finally the training of a body of young men capable of prosecuting research in pure science and, *ipso facto*, in so-called industrial research ... The nature of the research work which the student should do in this training should be ... in pure science, rather than in any problem directly connected with industry or looking to immediate practical application.[86]

On the basis of this conception of the relation between the universities and industrial research, council members proposed foundation of a national research institute, along the lines of the National Physical

Laboratory in Britain and the American National Bureau of Standards, to solve industrial research problems in concert with Canadian industry.[87]

By proposing that the training of researchers in pure science was a necessary and even sufficient condition for the development of industrial research, the NCCU and especially the NRC clearly favoured the position of university researchers, who, for many years, had insisted on the importance of scientific research. Moreover, this stance allowed the latter to translate their claims into student training which conformed to the traditional role of the university. In effect, by advancing their research project and (simultaneously) their position in the world of science, university researchers often gave the impression of pursuing two independent activities, teaching and research, the second more and more parasitic on the first. They could now invoke the importance of training researchers to convince their institutions that teaching and research were inseparable and that, henceforth, research should be funded in order to assure adequate training of students. We have here the origin of a discourse, often repeated since, which claims that only research can assure high-quality teaching at the university level.

In 1918, the annual report of the physics department of McGill University was already invoking the NRC's prestige to justify an increase in the resources allocated to research: "In view of the establishment of studentships under the Honorary and Advisory Council for Scientific and Industrial Research, there will be an increasing demand for graduate instruction and research, especially after the conclusion of peace. To meet the needs of research in the field of modern physics, a special annual income for the purchase of new instruments and laboratory supplies is urgently needed."[88]

While Toronto and McGill were the first institutions to emphasize scientific research, other Canadian universities quickly fell in line, pushed as much by the NRC's promotional discourse as by competition among Canadian universities.

In his report for the 1916–17 academic year, the president of the University of Saskatchewan, Walter C. Murray (a member of the NRC) noted: "The war has awakened the nations to the importance of scientific research." Recalling initiatives by the various governments of the empire and the United States in organizing research, he continued: "The nations of the world today have come to see what the scientists have long preached, that in science they have one of the most potent of instruments for extending human power."[89] As a discrete but symbolically significant sign of the new interest in research, the report included for the first time a list of the publications of professors, whose research had acquired greater visibility within the institution. The university thus imitated a practice inaugurated at the beginning of the century by Toronto and McGill.

At Queen's, the president's report for 1917–18 also contained a complete section on the importance of research, where it was noted: "Since the beginning of the war there has been a complete revolution in the British attitude towards this subject and government and people are now making haste to recover lost ground and to put the nation in a strong position in this respect."[90] The president concluded: "In order to do her part in the nurture of scientific research, Queen's must be well staffed and well equipped in all departments of science. It is particularly necessary that those professors who show aptitude for research and zeal in its pursuit should not be handicapped by long hours of teaching."[91]

This discourse was not new: it repeated what his colleague physicist A.L. Clark had been promoting in his annual reports for the previous five years. The only difference – and this was fundamental – was that the discourse was now official: repeated by the university president, it now had a different status, as the president spoke in the name of his institution. It all happened as if the circumstances of the war, by bringing about creation of the NRC – a type of institution that then was being created in most countries in the world – favoured the discourse of the generation of researchers. First reiterated in the reports of department heads, the discourse on the importance of research acquired greater legitimacy by being taken up at a higher level, in the reports of the presidents themselves. As we have already seen, this movement was accompanied by the listing of researchers' work in a special section of the annual report.

By the 1920s, scientific research was considered an integral part of the modern university. In a report presented to the Ontario Commission of Inquiry on University Finances in 1920, the president of the University of Toronto wrote: "One of the chief functions of a university is to extend knowledge and to train others who will extend knowledge ... The experience of the Great war and after have rendered unnecessary any extensive advocacy of the value of scientific research ... It was the application of the results of scientific research that contributed largely to the successful conduct of the war."[92] While the first part of the sentence only restated a position already advanced by James Loudon toward the end of the nineteenth century, the second part gave greater credibility to this discourse, which had just begun to be translated into practice, thanks in particular to the NRC's financial aid.

NATIONAL POST-GRADUATE FELLOWSHIPS

Although the administrators of the major universities had already recognized increasingly the importance of scientific research and of training scientists, the institutions' financial difficulties had hindered the translation of this support into tangible form. At Toronto, for example, the launching of the PHD program did not lead until 1916 to establishment of an adequate structure of financial assistance for post-graduate students. The introduction of post-graduate fellowships at that time was, according to J.C. McLennan, head of the physics department, a crucial development. McLennan had been arguing for such a scheme for many years, in the interests of securing a healthy future for research in his department; in 1916 he wrote to the university president, R.A. Falconer, that "it looks as if a new era is opening for the University and I look forward for happier days now."[93] There was reason for McLennan's optimism, all the more so because the NRC would also be instituting a scheme of post-graduate financial support just a few months later. Ultimately, the University of Toronto fellowships would be directed to the disciplines not covered by the NRC awards.

In order to make possible the training of a larger number of researchers across the country, the members of the NRC had to establish a scholarship program that encouraged Canadian university science graduates to pursue post-graduate studies in Canada rather than the United States. Accordingly, action was taken at the first meeting of the NRC, in December 1916.[94] The university-based members formed them-selves into a committee to study the operating principles of the 1851 Exhibition Scholarship and to make recommendations on Canada's particular needs.[95]

Two types of assistance were eventually established. "Studentships," valued at $600 for the first year and $750 for the second, were to be awarded to applicants entering on their scientific studies, normally at master's level. "Fellowships," valued at $1,000 the first year and $1,200 the second, were directed to doctoral students.[96] As was true of the 1851 Exhibition Scholarships, the awards were to be confined to students who had already shown "high promise of capacity of advancing science or its applications by original research."[97] Both pure and applied sciences were covered. The first awards were made in September 1917, less than a year after creation of the NRC.[98] Although seventy awards had been anticipated, wartime conditions limited the number to seven. Not until 1923, in fact, did the NRC reach the point of spending its entire annual budget of $120,000.

The NRC studentships and fellowships undoubtedly met the require-ments of universities such as McGill and Toronto, which were already capable of introducing undergraduate students to research methods, thereby enabling them to qualify for studentships. At a profound disadvantage, however, were those institutions that were less well equipped. There, students had little opportunity to participate in research as undergraduates and so could not demonstrate their "high promise of capacity for advancing science." The result was a circular situation, where a student had to have had experience in research before being considered qualified for training. To break the cycle, the NRC instituted in 1919 a new form of financial assistance. Worth $500, the bursary was intended to encourage able students to begin post-graduate study. Any students who showed during the year of the bursary, "distinct evidence of capacity for original research" would then qualify for a studentship.[99]

Analysis of the distribution of these different types of award shows that, had the bursaries not been offered, universities such as Dalhousie, Queen's, and those in the west would have been unable to benefit from any of the NRC awards. Studentships and fellowships were virtually the preserve of McGill and the University of Toronto. Of the seventy-eight studentships and fellowships awarded in physics from 1917 to 1938, for example, three went to Queen's students and one to the University of Manitoba. The University of Toronto, meanwhile, received fifty-six, and McGill twenty-eight. By contrast, of the 100 bursaries awarded in physics from 1920 to 1938, thirty-seven went to the smaller universities; even so, Toronto received thirty-one and McGill thirty-two.[100] Despite variations among disciplines, reflecting the unequal strength of activities among the various scientific departments, the overall predominance of these two universities (see Tables A4 and A5) was overwhelming.

This situation was inevitable not only because of their large endowments but also in view of the expressed opinion of A.B. Macallum, as NRC chairman, that these two universities should be made centres of post-graduate study for the graduates of all Canadian institutions. As early as June 1918 Macallum put this argument in a letter to his sometime Toronto colleague J.C. McLennan:

One of our great difficulties, in connection with studentships and fellowships, is going to be the places of tenure of these positions. Already three of our fellows have expressed a request to go to the American Universities, which the Council did not think wise to grant. When immediately after the war the number of studentships and fellowships may be increased to fifty, and, ultimately, to one hundred, the problem will become an acute one, and, in view of this, I am proposing that the universities of Toronto and McGill should establish Science Research Faculties, composed of staffs specially selected for research work and the guidance of graduates desirous of entering a scientific career.[101]

In the previous year, during the fourth meeting of the NCCU – held in Ottawa immediately following a meeting of the Royal Society of Canada – Macallum had joined with F.D. Adams, A.S. Mackenzie, and the future NRC president H.M. Tory to prompt passage of a resolution setting up a special committee chaired by Adams. This body was "to take up with the authorities of the larger Canadian universities the question of organizing jointly graduate work leading to the PHD degree, and that it reports the result at the next meeting of this conference."[102] Later meetings of the NCCU took the proposal so far as to discuss creation of a national post-graduate university. This scheme had no chance of succeeding, however, because of the existing universities' fierce competition to attract students and the fact that education came within provincial jurisdiction.[103]

In effect, while discussions went on, Toronto and McGill were steadily consolidating their ability to attract the majority of aspiring Canadian post-graduate students who did not wish to move to the United States. As Table A4 indicates, McGill and Toronto profited the most from the scholarships offered by the NRC. Because recipients were obliged to undertake research work at an institution "where the conditions are thoroughly suitable, and the accommodation ample, for such researchers,"[104] these two universities enjoyed a clear advantage, and, consequently, their output of science graduates, at both master's and doctoral levels, increased greatly from the early 1920s onward (Figure A2).[105]

Obviously, the NRC's decision to concentrate its financial aid at universities that were already well equipped for research was not welcomed by all. Queen's University, spurred by the physicist A.L. Clark, created an interdepartmental committee on scientific research. In its first report, appended to the university principal's report for 1916–17, the committee stated its guiding principle forcefully: "It is essential, if Queen's is to maintain her rank among Canadian universities and is to contribute her proper share to the advancement of knowledge and to the development of our national resources, that increased attention and support be given to the world of research. ... Very little help is to be expected [from the NRC], the committee continued, to establish research work."[106] It recommended that the university establish its own research council, charged with distributing grants to researchers and paying for the hiring of research assistants. The support of summer research was to be emphasized, in view of the problems encountered by professors trying to combine research with teaching during the regular academic year. The committee further suggested that Queen's offer its own $600 scholarships and $1,200 fellowships.[107]

Yet its plans were to prove unrealistic. Full implementation would have required an annual expenditure of $25,000 to $30,000. In its first year the university could provide only $10,000, and there was no guarantee of renewal. As a result, no financial awards were made. However, Queen's did receive, between 1917 and 1927, twenty-one awards from the NRC. In terms of numbers of awards it was thus placed immediately behind McGill and Toronto and ahead of the thirteen awards granted to the University of Saskatchewan. Queen's would retain this third-place status throughout the inter-war years.

For A.L. Clark, creation of the Queen's committee on scientific research was only the beginning. Appointed dean of applied science in 1919, Clark immediately suggested to G.Y. Chown, retiring as registrar and treasurer of the university, the endowment of a research professorship.[108] Chown agreed, and the Chown Research Professorship was created for the fields of chemistry or physics. The first incumbent was an English physicist, A.L. Hughes, who departed four years after his 1919 appointment to take up a position in the United States. He was succeeded by another physicist, the Australian J.A. Gray, who held the position until retirement in 1951. The establishment of this position had a marked effect on the expansion of research work in physics at Queen's. Between 1923 and 1939 the department's students gained fourteen NRC awards, three times as many as any other physics department except those of McGill and Toronto.

In summary, the post of researcher allowed Queen's to increase its chances of obtaining scholarships for its students. These scholarships, in turn, were essential for the development of research. Clark was perfectly conscious of this, and in 1927 he reiterated his position that "in order that an increasing number of those interested in graduate study may return for the master's degree, a system of fellowships tenable for one year would be most valuable."[109]

Hughes's departure in 1923 was not unrelated to the department's meagre resources. In March 1922, he explained to A.N. Shaw at McGill how discouraging it was not to be able to compete with researchers at the National Bureau of Standards in Washington, who were working on the same problem as he:

Foote and Moohler at the Bureau of Standards, Holweck in France, are methodically clearing up the subject. The former [Foote] told me at Christmas that he had as many as seven physicists (PHD's) working on his problems never less than three, besides the resources of the standards lab in material, apparatus, mechanics and glass blowers. It is rather discouraging to compete with such conditions. To think of a piece of apparatus and a method of carrying out a piece of research at the Bureau of Standards is almost equivalent to have it done! Here, in spite of lots of time, it is equivalent to months of dull, uninteresting preparatory work.[110]

The following year, he left Queen's to take up a position at St Louis University in Washington.

In the beginning, Gray also complained of a lack of collaborators. In a letter to Rutherford, he wrote in 1926: "I have only one research student at present. I have three x-ray outfits with a fourth one nearly complete and no one but myself to work them."[111] Happily for Gray, two of his students gained NRC awards in the following year, and by 1928 he was supervising three such award-holders.

For the members of the NRC the award of financial support to post-graduate students constituted only a first step toward establishment of a systematic research capacity in Canada. With post-graduate awards in place to encourage students to enter research, it was equally essential that professors should also be able to devote themselves to research activities. Therefore, as soon as the regulations for the post-graduate awards had been defined, the council set about designing a scheme for subventions to researchers.[112] At the third meeting of the council, W.C. Murray suggested creation of a committee to study the methods used by other organizations to help finance development of research projects. Under the direction of A.S. Mackenzie, a committee composed of Adams, McLennan, and Kirkpatrick went to work.[113]

After having looked at the workings of the American Academy of Arts and Sciences, the Carnegie Corporation, the Rumford Fund, and the Smithsonian Institute, as well as several other organizations, the committee proposed regulations for the awarding of grants that were immediately accepted by the council. The quality of the applications would be evaluated by members of the different NRC committees who would subsequently make recommendations to council. The grants, made initially for a single year but renewable on reapplication, would "as a general rule, only be made to persons who are conducting investigation in established laboratories which possess the fundamental apparatus and facilities necessary for research of the nature proposed, and ... will not be made for the purchase of standard apparatus which a well equipped laboratory should possess."[114] Naturally, this stipulation gave a further advantage to institutions already well equipped – notably McGill and Toronto – and gave further cause for grievance on the part of the universities that thought they would be effectively excluded from the scheme.[115]

The portion of the NRC's total budget allocated annually to different research projects grew continually until the end of the 1920s, by which time it had attained 65 per cent of the total budget (Table 2.3). The sharp fall in research expenditures from 1932 on can be explained by the Depression and by the opening that year of the NRC laboratories, which allowed the organization to do its own research and to spend a greater part of its budget internally.[116] Prior to 1932, problems considered important were referred directly to competent university

Table 2.3
NRC Budgets (Current Dollars), 1917–38

Fiscal Year	Scholarships Grants	Research Expenses	Total
1917–18	5,550	8,344	50,374
1918–19	7,150	19,511	61,151
1919–20	20,100	39,875	99,872
1920–21	16,925	60,779	109,578
1921–22	37,525	24,617	99,564
1922–23	34,975	39,511	106,369
1923–24	37,830	59,854	120,000
1924–25	4,011	50,972	120,000
1925–26	40,082	61,931	135,000
1926–27	40,105	74,464	150,000
1927–28	41,855	107,926	170,000
1928–29	43,720	193,977	300,000
1929–30	49,990	220,442	400,000
1930–31	59,535	178,923	500,000
1931–32	38,490	166,898	478,000
1932–33	17,600	15,820	410,000
1933–34	9,160	25,292	404,500
1934–35	11,825	32,943	392,000
1935–36	13,205	44,041	400,000
1936–37	15,675	66,146	500,000
1937–38	22,813	10,799*	637,800
1938–39	28,309	7,762*	758,680

Source: NRC, Annual Reports.
* These figures include only grants to individual researchers.

researchers. The opening of the NRC laboratories was done in the depths of the Depression, and the amounts allocated for scholarships and research grants had been greatly diminished, although the NRC's total budget had shrunk by only 20 per cent between 1930 and 1935.[117]

In addition to obtaining help from the NRC, some professors received grants from private foundations. Between 1911 and 1941, for example, the New York–based Carnegie Foundation made a significant contribution to research in the major Canadian universities (Table 2.4). Toronto received at least $135,000 during this period. McLennan, for example, used funds from this source to put together his low-temperature physics laboratory. McGill received only $4,800 for research but gained more in the form of development and endowment funds,

Table 2.4
Grants (us Dollars) by the Carnegie Foundation to Canadian Universities,
1911–41

University	Research	Development	Support and Endowments	Equipment
Acadia	5,000	26,200	275,000	22,500
Dalhousie	5,500	248,000	1,065,000	16,000
McGill	4,800	168,050	1,000,000	17,250
Queen's	12,000	267,800	100,000	22,500
Toronto	135,000	64,100	–	44,150
Saskatchewan	5,000	97,500	–	19,000
Alberta	13,000	106,000	–	22,500

Source: Robert M. Lester, *A Thirty Years Catalog of Grants*, Carnegie Corporation of
New York, 1942.
Note: Only universities that received research grants are included.

which contributed indirectly to the growth of research.

Another important source was the Rockefeller Foundation. McGill professors received from this organization for the sciences alone, excluding medicine, $86,596 between 1934 and 1944 for work in embryology, genetics, neurology, and pharmacology. John S. Foster, the physicist, received $2,100 in 1938 to study the possibility of constructing a cyclotron which was built after the war. Two University of Alberta professors received respectively $500 and $1,500 for research in organic chemistry and zoology. Between the two world wars, Canadian universities benefited especially from the Rockefeller programs in medicine, public health, and the social sciences.[118]

CONCLUSION

In 1927 the NRC summarized the first decade of its activities and was proud to announce that "an active and efficient research organization has been built in Canada, through which the investigation of any problem of national importance can be undertaken." NRC funds had permitted 199 students sharing 344 fellowships, studentships, and bursaries to be trained at the graduate level in twelve universities across Canada. Though "the main purpose of scholarships [was] to train men in research work," not necessarily to produce publications, the report noted that the 458 scientific papers produced by these scholarship-holders were a good indication of the seriousness of their work.

Sensitive to the problem of the "brain drain," the report mentioned that out of the 155 students who had completed their training, 123 were still in Canada. In its survey of the research grants, the council noted that more than 120 projects involving universities had been financed and that, although the majority of them involved specific industrial problems, the NRC had also financed pure research, on the grounds that "industry can advance no faster than the principles upon which it is based." As a further justification, the report stressed also that "it is not often easy to differentiate between researches in pure and those in applied science, since it is frequently found that the pure science of yesterday has an industrial application to-day."[119]

It may be estimated that professors' individual research projects accounted for approximately 20 per cent of the NRC's annual budget allocated to research.[120] In addition to its scholarship program, the grant-in-aid-for-research program of the NRC thus played a fundamental role in the development of university research by stimulating the constitution of more autonomous disciplinary groups.

H. Blair Neatby has suggested that "English-Canadian universities have experienced a revolution since the 1940's" and that "the impetus of this transformation has been the priority given to research."[121] In fact, this "revolution" had a long preparation and was rather "quiet." As for the institutionalization of research described in this chapter, it was the result of a complex set of circumstances. Perhaps economic and social development always underlie university growth, but they cannot in themselves assure that the agents who occupy university positions will succeed in imposing their vision of the institution. The increase in the number of agents with research-oriented interests among the professorial corps of Canadian universities could only give greater force to the idea of the university as the natural site for the advancement of knowledge. However, researchers might have spent a long time in the desert if historical circumstances had not given weight to their discourse and helped to translate it into concrete terms. The First World War heightened awareness, on a world-wide scale, of the importance of scientific research. Leading to the creation of institutions specially devoted to the development of scientific research (pure or applied), this conjuncture reinforced a group that was then in full growth but had few resources for its self-assigned task of producing knowledge and reproducing itself as a group.

The creation of grants for researchers and scholarships for students by the NRC thus led to a rapid expansion in research in Canadian universities, which I shall analyse in the following chapter.

CHAPTER THREE

Growth and Diversification of Research

In varying degrees, university research in most disciplines – with the exception of mathematics, which, prior to the Second World War, did not fall within the NRC's mandate – owed a good part of its development to the possibilty offered by the NRC of establishing research teams which permitted the reproduction of the discipline. While this debt cannot be demonstrated in detail for each discipline, it is possible to compare the relative positions of Canadian universities with regard to scientific research on the basis of grants received from the NRC by individual researchers. Within this general framework, detailed study of the development of research in physics will bring to light more general processes also at work in other disciplines, such as biology and chemistry.

DISTRIBUTION OF RESEARCH ACTIVITIES

In both "pure" and "applied" sciences, Canadian universities did not all profit equally from the research grants made available by the NRC from 1917 on. A first indicator of university research activity is given in Table A6, which shows, for most universities, the annual number of projects funded by the NRC. Many projects were very much applied in nature. Projects in agriculture and livestock management, for example, were undertaken by the prairie universities and were spread out over many years, which explains why these universities rivalled McGill and Toronto in terms of number of NRC-funded projects. Although useful, the number of projects is clearly not an adequate indicator and can be complemented with the total value of individual research grants.[1] In this respect (see Table 3.1), Toronto dominates, while McGill and the prairie universities are more or less equal. In fact, the difference between Toronto and McGill can be explained by the performance of the Toronto physics department, which alone received around $30,000.

These funds were not equally shared by different science departments.[2]

Table 3.1
Total Value of Research Grants to Researchers by University (All Disciplines Included), 1917–37

University	Total (Current Dollars)
Toronto	64,910
McGill	49,557
Alberta	49,201
Saskatchewan	44,630
Manitoba	34,603
British Columbia	28,144
Montréal	15,280
Queen's	11,500
Dalhousie	6,670

Source: NRC, List of Research Grants to Individual Applicants, NRC Archives, Ottawa.

As Table 3.2 shows, the disciplines privileged by the NRC were physics, biochemistry, and engineering. In Alberta, the biologist Robert Newton received $17,000 for his studies in the genetics of wheat rust and the physicist Robert Boyle $11,000 for his work on ultrasound. In Saskatchewan, the concentration of grants was even greater, since T. Thorvaldson alone received $36,000 (of the $44,000 received by all professors) for his work on the deterioration of cement. This problem was particularly worrisome for the western provinces, where the highly alkaline water had a deleterious effect on cement used for construction. In Manitoba, problems related to agriculture accounted for half of the grants received, the remainder being distributed among chemistry and engineering departments. In British Columbia, oceanography was the most developed area of research.

It may come as a surprise to see the Université de Montréal on the table, outdistancing even Queen's and Dalhousie. Of the $15,000 received during the period, frère Marie-Victorin received $9,650 for his work on the flora of Quebec and eastern Canada. The same year, Paul Riou, of the chemistry department, received $300 to study the crystalline form of sodium bicarbonate. Louis Bourgoin at the Ecole Polytechnique obtained $2,385 for the study of the catalytic action of ultra-violet rays. Finally, A.V. Wendling, also at the Polytechnique, received $2,000 in 1926 to study the dielectrical properties of different classes of porcelain.

Between 1923 and 1938, following the opening of the Faculty of Sciences of the Université de Montréal in 1920, a dozen francophones benefited from NRC bursaries to pursue higher education. Ten spe-

Table 3.2

NRC Grants (Current Dollars) to Researchers by Discipline, 1917–37

Agronomy	2,600
Bacteriology	7,659
Biochemistry	64,765
Biology	7,260
Biophysics	4,690
Botany	25,283
Chemistry	18,300
Entomology	2,135
Engineering	83,585
Medicine	1,400
Microbiology	1,500
Oceanography	1,000
Petrochemistry	1,350
Physics	155,548
Veterinary medicine	5,000
Zoology	8,000
Total	390,075

Sources: NRC, List; NRC, *Annual Report.*

cialized in chemistry; two chose physics. Three studied at Laval University, four at McGill, and five at Montréal, including one at the Polytechnique. In terms of number of bursaries (each student having received an average of two), Montréal received twelve, nine of which were in chemistry, while Laval received seven, all in chemistry or biochemistry. The importance of chemistry at Laval was the result of the Ecole supérieure de chimie, where research took place owing to the presence of professors from the catholic University of Fribourg, who had founded the school in 1920. These professors trained disciples who, after having obtained their doctorates in Europe during the 1930s, found positions as professors at the school and continued their research, thus forming the first contingent of francophone chemists.[3] During the period 1917–37, no Laval chemist received a grant from the NRC.

In the two francophone universities, the training of researchers was limited to chemistry and botany and was a marginal activity compared to undergraduate teaching. In fact, it was necessary to wait until the end of the Second World War for research to be undertaken in other departments, notably physics.

At Queen's University, science professors received only $11,500 during the period 1917–37, less than the total amount of internal funds

Table 3.3
Average Annual Value (Current Dollars) of NRC Grants to Canadian Universities
(All Disciplines Included) 1917–37

University	Research Grants	Scholarships	Total
McGill	2,477	9,305	11,782
Toronto	3,245	6,850	10,095
Saskatchewan	2,218	1,010	3,228
Alberta	2,460	300	2,760
Manitoba	1,730	405	2,135
Queen's	575	1,336	1,911
British Columbia	1,407	315	1,722
Montréal and Polytechnique	804	443	1,247
Dalhousie	333	457	790

Sources: NRC, List; NRC, Annual Reports.

administered by the research committee, which was between three and four thousand dollars a year. In 1923, the committee provided $2,295 for research and a further $550 to pay several research assistants $70 a month.[4] In 1931, the amounts were $2,000 for research and $1,783 for assistants.[5]

To understand the interaction between these research activities and the training of researchers, it is necessary to relate the amounts received as grants with the amounts received in the form of student bursaries at each university. In my opinion, this is the best indicator of the level of development of university research, since production and reproduction should go hand in hand. As Table 3.3 indicates, Toronto and McGill dominate the Canadian university field, the amount in grants being in close agreement with the development of graduate education in these departments. The University of Alberta, however, which in terms of grants compared favourably with McGill, received few bursaries, a sign of the lack of development of graduate education in that institution. Only the University of Saskatchewan obtained a significant number of bursaries, with award-holders specializing in agriculture.

Overall, Canadian universities may be divided into three groups. McGill and Toronto make up the first, followed by Queen's and Saskatchewan, where the number of bursaries received is a good indicator of research. They are followed by Dalhousie, Manitoba, Alberta, British Columbia, Montréal, and Laval, where research was relatively underdeveloped. Although the situation might vary according to discipline – this, we shall shortly see, was the case with physics – this classification represents well the relative position of Canadian universities in terms of inter-war research.

As Figure A2 shows, the production of theses in physics as well as the number of papers published grew considerably after the First World War. Between 1900 and 1916, only thirty-three master's and six doctorates were granted in physics, while during the period 1917–39, these numbers rose respectively to 208 and eight-five. Although the doctorates were still granted by McGill and Toronto, a sizeable proportion of the master's degrees (42 per cent) was granted by Dalhousie, Saskatchewan, and Queen's. The proportion of publications produced by these universities also increased appreciably, growing from 10 per cent of the total for the period 1900–17 to 23 per cent in the following decade and 38 per cent in the 1930s (Figure A3).

After having obtained master's degrees, some of these students went to McGill or Toronto or even to the United States or Europe for their doctorate. In the Toronto and McGill physics departments, approximately a third of those who completed their master's went on to do a doctorate in the same institution. However, these students accounted for 40 per cent of the doctorates at McGill and only 23 per cent of the doctorates at Toronto, which confirms that Toronto had greater visibility in physics and could thus attract students from other Canadian universities more easily.

Between the two world wars, the Toronto physics department was without doubt one of the most important in North America. Reasearch bore mainly on atomic spectroscopy, superconductivity, and low-temperature physics, all of which activity was directed by McLennan.[6] Always abreast of the most recent discoveries, McLennan knew how to reorient his research rapidly in order to stay on the front line of research. Already in 1910, he became interested in spectroscopy and left the study of radioactivity, an area in which, with his colleague E.F. Burton, he had already made important discoveries.[7] During the war, he was attached to the Royal Navy as a scientific adviser. Examining the potential helium resources for balloons, he supervised the setting up of a production plant in Calgary. The war over, he pressed the importance of low-temperature physics; using the resources at his disposal at the Board of Invention and Research of the Royal Navy, he submitted a plan for the development of helium production to the university's president, Robert Falconer, saying that "with the research facilities in spectroscopy which I developed in Toronto with the support of the University, we are almost in a unique position in that field ... in America and in fact in the British Empire to deal with all problems associated with low temperature work."[8] In 1920, he wrote that "the successful production of liquid Helium would at once open up a large field of investigation to the physicists, chemists, physiologists, zoologists, and botanists of Canada, for it would enable us to study the properties of materials including chemical reactions of the life of bacteria, spores, etc. at the lowest temperature available."[9] After several years of work with his assistant, Gordon M. Shrum, then a doctoral student, he succeeded in early 1923 in producing large quantities of liquid helium, which had not been done since Kammerlingh Onnes had made some in his laboratory in Leyden in 1908.

Armed with this new equipment, the Toronto group was able to do pioneering work in the study of low temperatures. Only several months after having produced liquid helium, McLennan conceived of a project to solidify helium and to study the spectra of hydrogen, helium, and nitrogen at low temperatures. His objective was to solve the enigma of the aurora borealis spectra, which, he wrote, "from a scientific point of view ... would be of the highest importance." His hypothesis was that "the aurora spectrum [is] a degenerate form of the Nitrogen band spectra."[10] In fact, after several months of hard work, Shrum finally showed, contrary to widely held opinion, that the spectrum was produced by oxygen, not nitrogen.[11] This discovery, consigned with McLennan, earned the latter the Gold Medal of the Royal Society of London in 1928.[12] Among the important discoveries made at Toronto before McLennan's departure in 1932 were the observation of superfluid helium[13] and the confirmation of the existence of ortho-hydrogen and para-hydrogen phases in liquid hydrogen, which at low temperatures behaves as a mixture of gases.[14]

As a member of the NRC, McLennan vigorously defended his students' candidacies for bursaries and almost always managed to secure them fellowships ($1,000) instead of studentships ($600). This did not always please the chairman, A.B. Macallum, who did not consider all of McLennan's students to be as brilliant as McLennan believed them to be.[15] Between 1918 and 1932, the year he retired, he obtained half of all the bursaries granted in physics by the NRC. Moreover, for his own projects, he received a total of $25,000 in NRC grants – more than all the McGill professors combined (Table 3.4). He directed twenty-five of the twenty-seven physics doctorates awarded in this period as well as most of the master's theses and, with his students, wrote the majority of articles published by the department – on average, a dozen a year. Between 1926 and 1932, the mean number of articles rose to twenty-four a year, of which many were published in the *Transactions* of the Royal Society

Table 3.4
Total Value (Current Dollars) of NRC Grants in Physics by University, 1917–37

University	Total
Toronto	30,500
McGill	20,632
Alberta	16,701
Manitoba	7,795
Queen's	3,395
Dalhousie	2,300

Source: NRC, List.

of Canada. McLennan's great productivity caused his friend Ernest Rutherford to comment: "You must cost a lot of printing to your Society."[16]

Throughout his long career, McLennan was very active in international physics. Named fellow of the Royal Society of London in 1915 and knighted in 1935, he was among the thirty most-cited physicists in the world during the 1920s.[17] If the success of the Toronto group owed much to McLennan's entrepreneurial talent and enthusiasm, it was also related to the nature of the research he undertook. The study of atomic spectra allowed for a very refined division of labour, which facilitated project definition. This favoured the production of theses within a relatively short time. McLennan's authoritarian direction of students did not always receive the approbation of Macallum, who believed that "students cannot be made into independent researchers by being given a piece of work and guided through it every step by the head of the department."[18]

McLennan's colleagues had their own areas of research, but, compared to him, they had few resources. E.F. Burton studied colloids and collaborated on several projects with McLennan. Along with McLennan, he was the only one to receive NRC grants, but on projects quite distant from his own research interests.[19] After having studied the properties of x-rays, L. Gilchrist turned toward geophysics and mining exploration methods. After having participated in McLennan's liquefaction-of-helium project, H.A. MacTaggart studied the properties of thin films and the behaviour of gases in solution. Finally, J. Satterly examined the viscosity of gases and the surface tension of liquids.

McLennan garnered most of the master's and doctoral students. The other professors had little help in advancing their research projects.[20] Following McLennan's departure, Burton became head of the department, which continued its work in low-temperature physics.

Like McLennan, John Stuart Foster at McGill started a systematic research program. Following his arrival in 1924, there was a notable increase in the number of theses produced. In the context of his research on the Stark effect (the influence of a magnetic field on the behaviour of spectral lines), between 1924 and 1939, Foster directed nineteen theses, of which fifteen were at the doctoral level. All were devoted to the Stark effect. Alone, he supervised a third of the total number of theses and half of the doctorates awarded in this period. By 1925, the number of NRC bursaries received by the department was comparable to Toronto's.

By the mid-1930s, spectroscopy, Foster's specialty, began to lose ground. Following the discovery of the neutron in 1932, more and more physicists became interested in nuclear physics. The Danish physicist Niels Bohr, who had encouraged Foster in his research on the Stark effect, suggested that he switch to nuclear physics. Taking advantage of a rearrangement of departmental research priorities, Foster conceived of a particle-accelerator project.[21] Held up by the Second World War, Foster's projects in this area were realized in 1944 with construction of a 100 MeV cyclotron.[22] This instrument enabled Foster to continue to supervise a large number of theses, this time in nuclear physics, where the "industrial" production of degrees was even more marked.[23]

Unlike the Toronto physics department, which, under McLennan's iron rule, concentrated on spectroscopy, McGill cultivated a larger number of research areas, though not all were pursued as systematically as Foster's and McLennan's. A.S. Eve, for example, directed twenty-six theses on a variety of subjects, most of the time as co-director. Between 1926 and 1933 he received close to $3,000 from the NRC to study respectively the Raman effect and the magnetic susceptibility of alkaline metals, unrelated areas of research. After having examined electrical discharges in gases, D.A. Keys concentrated on geophysics, an area that also interested Eve. McGill also harboured Louis V. King who, at the time, was the only Canadian mathematical physicist of any importance. He worked on problems of electromagnetism and theoretical acoustics. In 1931, he was joined by the British theoretician W.H. Watson, who was also interested in electromagnetic theory. Altogether, of the sixty-one theses produced between 1918 and 1939, six were on radioactivity, nineteen on the Stark effect, two on quantum theory, and thirty-four on a variety of topics from the Raman effect to geophysics.[24]

Even though they did not have the same resources as their colleagues at McGill and Toronto (see Table 3.5), physicists at the prairie universities, at Dalhousie, and at Queen's still did research.[25] In Alberta, Robert W. Boyle's work on the properties of ultra-sound was well funded by

Table 3.5
Average Annual Value (Current Dollars) of NRC Grants in Physics to Canadian
Universities, 1917–37

University	Research Grants	Scholarships	Total
Toronto	1,525	4,247	5,772
McGill	1,031	2,247	3,278
Alberta	835	157	992
Queen's	169	436	605
Manitoba	389	168	557
Dalhousie	115	215	330

Sources: NRC, List; NRC, Annual Reports.

the NRC, which granted him $11,000 between 1917 and 1929. Introduced to research by Rutherford when at McGill, Boyle had worked on the properties of radium emanations until 1914. During the war, as a member of the British Board of Invention and Research, he participated in work on the detection of submarines using high-frequency waves. He continued this work after the war, thus abandoning radioactivity. Before leaving his position in Alberta in 1929 to join the NRC as director of the physics division, he published fifteen articles, most of which appeared in the *Transactions* of the Royal Society of Canada.[26]

Research in his department diversified with the arrival of S. Smith in 1920 and R.J. Lang in 1924. Both were active in spectroscopy, Lang having received his doctorate from the University of Toronto the previous year. Between 1925 and 1934, they jointly received $5,700 for a project on atomic spectroscopy in the ultra-violet region that produced about two publications a year. Nonetheless, the department received only five NRC scholarships, and the award-holders all worked on Boyle's research program. The department does not seem to have awarded more than fifteen master's degrees in the period under consideration.

In Manitoba, the situation was quite similar. Not more than ten master's degrees were awarded, and only five scholarships were received from the NRC. All the award-holders worked on projects related to Frank Allen's work in biophysics. Alone or with his students, Allen published eighteen papers, and three of his students published alone, which made a total of twenty-one papers between 1923 and 1931, or an average of two a year. From 1912, the department had harboured R.K. McClung who, ten years earlier, had worked at McGill under Rutherford.[27] At Manitoba, he abondoned research and devoted himself to teaching.

In 1924, with the arrival of one of McLennan's students, J.F.T. Young, the department widened its research to include spectroscopy. In 1924, E.W. Samson received a scholarship from the NRC to do a thesis in this area under Young's direction.

Even though Queen's University's physics department, directed by A.L. Clark, received fewer grants than Alberta or Manitoba, a number of research projects were developed between the wars with the help of the research committee formed in 1916. Moreover, since 1919, Queen's had had a full-time research position. Even though most of the professors had research projects, none had the time to make quick progress. It was for this reason that the research committee emphasized the hiring of research assistants during the summer months. Helped by a total of ten students, Clark, who from 1919 was dean of the Science Faculty, was able to advance his research on the thermodynamic properties of liquids and gases and published a dozen papers between 1918 and 1939.[28] As for J.K. Robertson, who joined the department in 1915, his research bore on spectroscopy and optics and occasioned twenty publications in the same period.[29] Finally, his colleague William C. Baker studied the adhesive properties of mercury and published ten papers. In spite of all this research, it was the task of the Chown Research Professor to direct master's and doctoral students.

The first holder of this position, A.L. Hughes, directed few students, the first NRC scholarship being awarded on his departure. The award-holder, Percy Lowe, who had been introduced to spectroscopy by McLennan,[30] was also awarded Queen's University's first doctorate in physics, with a thesis entitled "Intensity Relations in the Spectra of Gases and Their Bearing on Atomic Theory." Once he received his degree, he secured a position at Kingston's Royal Military College. Hughes's successor, J.A. Gray, also had a difficult beginning. Attracted by the possibilty of devoting himself full-time to research, he had left an associate professorship at McGill. As we saw in the previous chapter, he had complained of a lack of collaborators, a resource that Hughes had also found lacking and which had caused his departure. Luckily for Gray, the number of students in physics grew rapidly from 1927 on.[31] He also took advantage of the help of fifteen assistants paid from the research committee's fund. By the eve of the Second World War, he had directed almost all the theses produced by the department, approximately fifteen (of which twelve had benefited from NRC scholarships).

While most physics departments in Canada concentrated on research in spectroscopy, Gray continued in atomic physics. In the mid-1930s, this area became nuclear physics, and Gray tried to convince university authorities to build an electron accelerator, but this was beyond the finances of the institution. He had to wait until the end of the war to obtain part of the necessary funding ($200,000) from the Atomic Energy Control Board, which, in 1947, also financed the construction of accelerators for McGill, Saskatchewan and British Columbia.[32] In 1950, the 70-MeV electron accelerator came on line, but Gray, then sixty-six, retired two years later. During his career, he directed fifty students. Before 1950, when the McGill cyclotron began to bear fruit, Gray had trained most of the nuclear physicists in Canada.[33]

While Queen's University, which received few grants, was able to count on the support of NRC scholarships to develop research in physics, Dalhousie was forced to rely on its own resources. With only five NRC bursary-holders (two in the early 1920s and three in the early 1930s), the department none the less produced thirty-two master's theses. This was 50 per cent more than Queen's, where the majority of graduates were award-holders.

When H.L. Bronson, a Yale graduate (PHD, 1904) and Rutherford's collaborator at McGill, began his university career in 1910, the department gave him only one assistant. In 1911, for example, he was forced to teach seven courses and direct nineteen hours of laboratory tutorials. The only professor until 1919, he directed three master's students. The first two, J.H.L. Johnstone (MSC, 1914) and G.H. Henderson (MSC, 1916), after having been introduced to thorium research, obtained their doctorates respectively from Yale (under the direction of B.B. Boltwood in 1916) and Cambridge (under Rutherford's direction), both owing to the 1851 Exhibition Scholarship. They returned to Dalhousie to teach alongside their mentor (the first in 1919 and the second in 1924). Bronson's first female student, Merle Colpitt, became his wife and, like Harriet Brooks twenty years earlier, left research.[34]

Toward the end of the First World War, Bronson left radioactivity to devote himself to the study of the specific heat of metals, which gave rise to seven articles published in collaboration with his students. With his colleague and former student G.H. Henderson, he directed most of the theses produced by the department. Until the mid-1930s, Henderson continued his work on alpha particles, but then he turned to geophysics, using radioactivity dating techniques on minerals. His work in this area earned him fellowship in the Royal Society of London in 1942.[35]

Henderson was the only member of the department to have received NRC funding. In collaboration with the chemist D. McIntosh, who had returned to his native province in 1922, he received $2,300 to study the therapeutic applications of radium radiation. The fifteen articles Henderson published between 1924 and the beginning of the war counted for more than half of the department's production, which was only twenty-four articles.

If it is true that, in general, NRC grants and scholarships played a determinant role in the growth of physics as a discipline in Canada, it does not follow that departments that benefited little or not at all from this source did not do research. We have just seen that this was not the case with Dalhousie, and Saskatchewan offers yet another example.[36] At the end of the First World War, the physics department there was starting from zero, so to speak. J.L. Hogg had just left, and a new physics and biology building was under construction. A.G. McGougan arrived for the 1919 fall session. He had to recruit other professors and prepare the new office space.

In the fall of 1920, he was joined by E.L. Harrington, a Chicago graduate, who had received his doctorate in 1916 for having measured the viscosity of air. Carried out under the direction of R.A. Millikan, this measure had been important for the determination of the value of the fundamental electric charge.[37]

Interested in instrumentation, Harrington specialized in the construction of new apparatus. He also directed research in the absorption of x-rays and, in the mid-1930s, became interested in nuclear physics and in the measurement of absorption cross-sections of neutrons and in disintegration through slow neutron bombardment.[38] This research allowed his five graduate students to write their master's theses.

According to Balfour Currie – a student there from 1927, and professor from 1930 – the arrival of Thomas Alty in 1925 marked the beginning of sustained interest in research.[39] Although A.E. Hennings had published several papers on the photoelectric properties of metals during his stay at the department – from 1914 to 1918 – it was not until the mid-1920s that research there integrated graduate student training. Like Dalhousie, the department had only five NRC bursary students, although twenty-eight master's degrees were awarded between 1927 and 1939. During his ten years in the department, Alty directed eleven theses, devoted to the study of surfaces.

When he left in 1935 to take up a position at the University of Glasgow, he was replaced by the spectroscopist and future Nobel Prize-winner Gerhard Herzberg.[40] The latter, too, stayed ten years and directed ten graduate students in spectroscopy. After two years at the Yerkes Observatory at the University of Chicago, he returned to Canada to pursue his research in the NRC laboratories. Alty's first student, Balfour Currie, returned to the department in 1938 after having obtained his PHD in meteorology at McGill. A new research program thus began, and, on the eve of the war, four students chose to devote themselves to the study of the aurora borealis under his direction.

In spite of all this activity, manifested in an increase in the yearly number of publications, the professors in the department did not call upon the NRC for help. Their first grants were received in the period

1938–39. In 1937, Herzberg received $1,300 from the American Philosophical Society for his study of the solar spectrum which had occasioned the purchase of a large-scale diffraction grating. As Harrington noted in his report, the grant made possible "a number of investigations which have been planned by Dr. Herzberg and are to be made under his direction by our students working for the master's degree."[41] It should be noted that the growth in departmental research coincided with the Depression, which led to a decrease in the number of NRC bursaries and the size of NRC grants.

Although the economic and social crisis of the 1930s led to a reduction in professors' salaries and university revenues, it did not lead to any sizeable decrease in the number of master's and doctoral theses produced, even though the NRC's budget for scholarships fell from $60,000 in 1930–31 to $17,600 in 1932–33. Overall, student enrolment in Canadian universities remained the same. In some universities, particularly in the sciences, it rose. At Queen's, for example, the total number of students fell slightly (from 931 in 1929–30 to 846 in 1934–35), while the number of students in science and medicine rose (respectively, from 404 to 454 and from 296 to 312).[42] The sharpest drop was among women (from 404 to 347), which led the dean of women, Hilda Heard, to comment: "A young man unable to find employment is sent to a University, a young woman is kept with domestic work."[43]

The same remarks applied to the few women who had professorial positions. According to Elizabeth Allin, who, after having obtained her doctorate in physics under McLennan's direction in 1931, became lecturer at the University of Toronto, "at one point all University staff members were asked to accept a reduction in salary, and married women in all departments were under pressure to resign."[44] According to Balfour Currie, the growth in university enrolment led to the hiring not of new professors but of a larger number of course assistants, often chosen from among the graduate students. He added: "The pay was modest but it was sufficient to induce students to elect special courses in physics. For most of them, the monthly wage made the difference between a frugal existence and one of some comfort."[45]

In the francophone universities in Quebec, we have seen that research – especially in the natural sciences and chemistry – had a rather modest position among professorial activities in the 1920s. Research in physics really began on the eve of the Second World War. In 1939, Laval hired the Italian physicist Franco Rasetti, who modernized the department and trained the first researchers. Before leaving for Johns Hopkins University in 1947, he directed several theses in nuclear physics and cosmic rays.[46] From 1947 to 1950, the department was directed by another Italian physicist, Enrico Persico, who specialized in electronic spectroscopy. The first generation of French-Canadian physicists trained at Laval acquired their autonomy, and in 1950 the direction of the department was handed over to Henri-Paul Koenig, one of Rasetti's first students.

In 1945, it was the turn of the Université de Montréal to modernize its teaching in physics and to initiate research by hiring the French physicist Marcel Rouault. Originally specialized in the x-ray analysis of molecular structures, he became interested in the physics of cosmic rays while at Montréal and constructed a small Wilson cloud chamber, transportable by balloon.[47] In 1947, he was joined by Pierre Demers and Paul Lorrain. The former, trained in Paris in the Joliot-Curie laboratory, continued the research on the use of photographic emulsions for the detection of high-energy nuclear reactions that he had begun during the war at the NRC laboratories, then located at the Université de Montréal.[48] The latter, a McGill graduate, undertook the construction of a Kockroft-Walton 500 KeV proton accelerator.[49]

In spite of the Depression, the number of university positions in physics grew throughout the 1930s, especially in the second half of the decade (Figure A1). In 1928, there were forty-six professors dispersed throughout Canadian universities. This number had grown by 1934 to fifty-seven and by 1940 to sixty-five. This increase varied greatly among universities, with Toronto accounting for almost half of the new positions, while the smaller universities, such as Acadia and Mount Allison, lost positions.

Compared to 1928, the general character of the population of physics professors had changed little in terms of the origins of their degrees. Almost half of the professors had then obtained their highest degree (usually a doctorate) in Canada, one-third in the United States, and 20 per cent in Britain. Twelve years later, half of the professors still held a doctorate from a Canadian university, 25 per cent from an American university, and 20 per cent from the United Kingdom.[50]

During the period 1918–39, McGill and Toronto and their physics departments dominated Canadian university physics. However, the growth of the prairie universities, Queen's, and Dalhousie offered graduates of McGill and Toronto the possibility of employment in a university where they would be able to continue research begun during their doctoral years. Thus J.F.T. Young in Manitoba, R.J. Lang in Alberta, and A.B. Mclay at McMaster (which moved to Hamilton in 1931), all students of McLennan, were able to continue in spectroscopy. After their stay at Cambridge, G.H. Henderson and D.A. Keys returned to Dalhousie and McGill respectively where they in turn directed students' research.

The presence of all these young researchers allowed many physics departments to diversify their research. Thus, at the end of the 1930s, Canadian physicists were working in geophysics, low-temperature physics, meteorology, nuclear physics, and spectroscopy. Thirty years earlier, research at Toronto and McGill had been concentrated on radioactivity.

The vast bulk of Canadian physicists' research was experimental. Although McGill had for a while harboured several theoreticians (King and Watson in physics and A.H.S. Gillson in mathematics), they did not produce any "offspring". At Toronto, McLennan was conscious of this gap in his department. In 1926, he suggested to President Falconer that a chair of theoretical physics be established in order to create "a graduate school of theoretical physicists paralleling what we have already done in the field of experimental physics."[51] In the absence of local theoreticians, McLennan took advantage of European physicists' visits to the United States and invited them to give series of courses in the department. In 1928, for example, H.A. Kramers gave a complete course on Heisenberg's new matrix mechanics and Schrödinger's wave mechanics. Wishing to call Falconer's attention to the importance of theoretical physics, McLennan took advantage of the occasion to write to Falconer, explaining that: "Professor Kramer's lectures ... have been epoch making as far as our department is concerned. It was really a wonderful presentation of the new wave mechanics."[52]

In spite of McLennan's pressure tactics, the chair in theoretical physics never saw the light of day. To his great displeasure, a department of applied mathematics was created in 1930, in order to attract John L. Synge, who had left the mathematics department in 1926 to take up a position in Dublin.[53] The theoretical physicist Leopold Infeld worked in this department from 1938 to 1950. Synge and Infeld were especially interested in general relativity.[54] At McGill, Foster compensated for the absence of theoreticians with a sustained correspondence with Niels Bohr, with whom he discussed the interpretation of some of his results.[55] In fact, theoretical physics did not enter physics departments in Canadian universities until after the war, and in direct relation to the development of nuclear physics.[56]

CONCLUSION

The post–Second World War period saw rapid growth in the number of graduates at the master's and doctoral levels. As W.P. Thompson noted in his study of the development of higher education in the sciences in Canada: "The launching of PHD programmes by so many universities after the war reflects the growth in resources and in number of students, their increasing interest stimulated by the war in research and graduate work, their recognition of its importance to the nation, and their realization that work at the doctoral level must be carried on in order to recruit and retain the best staff."[57]

To extend our study of the development of research in physics would lead essentially to a chronology of research. The inter-war period was the time when a system of university research was established and all physics departments recognized research as part of their function. It is true that afterward new areas of research emerged, but research itself was now taken for granted and quantitative growth led to few qualitative changes.[58]

Between 1900 and 1939, in contrast, quantitative growth led to qualitative change. The emergence and institutionalization of research within the universities affected other institutions such as the Royal Society of Canada, which was forced to respond to the new needs of researchers. In particular, expansion of research, especially in physics and chemistry, raised the issue of diffusion of research results. This problem was solved by the NRC's creation in 1929 of the *Canadian Journal of Research* (*CJR*). The NRC's Associate Committee on Physics and Engineering Physics, formed in 1919, became the first organ to speak for physics as a discipline. The increase in the number of physics positions in governmental institutions, industry, NRC laboratories, and provincial research organizations during the 1930s and 1940s gave rise to a new series of demands, different from those advanced by university physicists during the 1920s. In response, there emerged first the Canadian Association of Professional Physicists (CAPP), and then the Canadian Association of Physicists (CAP), which represented physics as a discipline in the period after the Second World War.

The history of these transformations, which took place simultaneously with – and helped make possible – those changes so far described, is the subject of the rest of this book.

PART TWO

Reforming Institutions

CHAPTER FOUR

Adjusting the Royal Society of Canada

The first two decades of this century saw the establishment of institutional structures within Canadian universities that allowed researchers to do research on a regular basis and to train new recruits. This development, already examined, made possible the existence of physicists as a group. However, exchanging and diffusing of research results were required if physicists were to contribute to the advancement of knowledge in physics and to obtain international recognition for their contributions.

Thus at the same time as they were establishing the means of production within their universities, the first Canadian physicists sought also to modify the Royal Society of Canada and, in particular, its *Transactions*, in order to make the society and its publication a tool for intervention within the field of physics.

A MEETING PLACE FOR PROFESSORS

As we saw in chapter 1, the Royal Society of Canada, founded in 1882, was intended to bring together the most eminent representatives of the main scientific and literary disciplines. However, given the level of development of Canadian university institutions at the time, the rather restricted number of potential members was such that the élite character of the society was manifested less in the exceptional productivity of its members than in the limited number of those who could actually join.

We also saw that the activities of the generation of professors active before the 1880s resembled more those of teachers than reseachers. For this reason, until the turn of the century, the meetings of section III (comprising chemists, physicists, astronomers, and mathematicians) differed little from those of the literary sections. The members of the different sections met for several days in Ottawa each year, in the third week of May, to present papers and to discuss among themselves. A year later, the *Proceedings* and *Transactions* of the society were published, and they contained, in addition to minutes of sessions, revised versions of some of the papers presented by the members. Until 1900, about twelve members met in section III. Each presented on average one paper a year; eight papers were published each year in the society's *Transactions*. The authors did not seem upset that their work was published so long after presentation. In other words, the members of the scientific sections did not produce for a market where competition demanded fast publication of results.

Among the physicists, only J.G. MacGregor, whose papers referred explicitly to the work of other European physicists and who produced for this market, might have had reason to complain, but the minutes of the meetings of the Royal Society contain no trace of his having done so. He seems to have solved the problem by publishing some of his papers simultaneously in Canada and England.[1] As long as most members of the scientific sections considered their papers to have a local character – and to be addressed mainly to members of the society rather than to an international scientific community, characterized by competition between agents seeking to maximize the material and symbolic profits of their work – there was nothing to suggest that publication should be speeded up.[2]

Once printed, the volume containing history, literature, biology, chemistry, and physics papers was distributed to other learned societies around the world, which in turn supplied the society in Ottawa with their own transactions. The society's activities – the annual meeting and publication of the *Transactions* – were made possible by an annual grant of $5,000 from the dominion government. The *Transactions* were also distributed to the lieutenant-governors of the provinces, the premiers, the ministers of education, and Canadian universities. In 1894, for example, 1,500 copies of that year's *Transactions* were printed.

THE INSTITUTIONAL IMPACT OF
A SCIENTIFIC REVOLUTION

At the turn of the century, the election of research physicists active in the field of physics, such as Ernest Rutherford (elected in 1900), Howard T. Barnes (1902), and John C. McLennan (1903), called into question the traditional modus operandi of the Royal Society of Canada. Arriving at McGill in 1898, Rutherford had brought with him an entire research program on radioactivity – a phenomenon discovered two years earlier by the French scientist H. Becquerel – and his many discoveries were, as we know, the basis of major conceptual transformations.[3] Following a stay at the Cavendish Laboratory, where he worked under J.J. Thomson,

J.C. McLennan (four years younger than Rutherford), returned to Toronto and surrounded himself with a team of dynamic collaborators able to compete with Rutherford in the area of radioactivity. Both groups participated regularly in the annual meetings of the American Physical Society, sometimes presenting identical results.[4] A disciple of Hugh Callendar, H.T. Barnes was also an active researcher in a less spectacular area, thermometry.[5]

At the annual meetings of the Royal Society of Canada, these physicists incorporated a more competitive conception of scientific practice, which required a society adapted to the dynamic of the scientific field in which they participated and in which they had been trained to function. Thus in 1904, Rutherford became vice-president of section III and as such joined the council of the society. In their annual report, council members raised for the first time the problem of publication of the society's *Transactions*. They pointed out that it was difficult to publish quickly a volume that contained contributions from different disciplines and that, given these circumstances, publication delays were frequent. This procedure hindered the rapid publication of important results, and as noted in the council report: "Delay in the announcement of a scientific discovery may be serious to original investigators, and, therefore, will not be sent to our volume of *Transactions* for publication."[6] The society should therefore adapt to the rapid change within the scientific community. Referring implicitly to Rutherford and McLennan's work on radioactivity, the report spirited in the society's *Proceedings*, stated:

The revolution in scientific thought now in progress is fundamental, and some of our members are in the van of the movement. Conceptions of the constitution of matter which have been held for ages, even in recent times, are being profoundly affected. Under such conditions, and they have arisen very suddenly and recently ... it might be well to inquire whether it should not be advisable to meet the emergency by issuing a bulletin ... In this way priority of discovery can be secured, and separate papers might be issued from the bulletin type.[7]

A committee presided over by A. Johnson of McGill therefore reported to the general meeting, which in turn adopted a proposition, seconded by McLennan, to the effect that the rules relative to publication of the *Transactions* be modified in order to allow immediate publication of work deemed important by the publication committee of the section concerned. Once accepted, an article would be sent immediately to the printer, "with the date of reception marked prominently."[8]

The first such bulletin was printed in June 1907. It contained two studies of the properties of radium and its emanation (radon) carried out by A.S. Eve, in collaboration with the chemist D. McIntosh, under Rutherford's direction.[9] Ironically, it appeared one month after the departure for Manchester of Rutherford, who had been at the origin of this new system of publication. The following year, it was McLennan's turn: a bulletin quickly publicized the results of his research "On the Radioactivity of Potassium and Other Alkalai Metals." These results were also published several months later in the British journal *Philosophical Magazine.*[10]

The society's awakening happened at the same time as Rutherford acquired an international reputation. In the spring of 1904, he was invited to give the prestigious Bakerian Lecture at the Royal Society of London, from which he received several months later the Rumford Medal. His book *Radio-Activity* appeared at the end of the year, and, at the beginning of 1905, he was invited to give the Silliman Lectures at Yale, which resulted in publication the following year of a second book, *Radioactive Transformations.*[11]

The spectacular nature and public impact of Rutherford's discoveries impressed the fellows of the Royal Society of Canada and made them aware of the value to the society of having these types of contributions published in its *Transactions.* However, in the long run, only increased numbers and representation within the society would allow research scientists to ask for and obtain further change in the publication of the *Transactions,* with a view toward producing a journal adapted to the new market in which researchers circulated their findings. Issuing bulletins to accelerate publication of important findings did not make the *Transactions* a journal like the *Physical Review,* which, since 1893, had been published on a regular basis and was specifically addressed to physicists.[12]

An efficient means for Rutherford, McLennan, and Barnes to maintain their position within the Royal Society was to make sure that individuals who shared their vision of scientific activity were elected members. Before leaving McGill therefore, Rutherford made sure that his seat in the society would be taken over by one of his peers. Apart from McLennan, the only physicist then active in research in Canada was Arthur Stanley Mackenzie, holder of the Monro Chair of Physics at Dalhousie University since 1905. After graduating from Dalhousie in 1885, he had been J.G. MacGregor's teaching assistant and in 1889 had enrolled at Johns Hopkins, where, under Henry Rowland's direction, he had written a thesis, "The Attraction of Crystalline and Isotropic Masses at Small Distances," for which he received his PHD in 1894.[13] After having taught for ten years at Bryn Mawr, he was offered a position at Dalhousie.[14]

Before taking up his new post, Mackenzie used a sabbatical year to recycle himself at the Cavendish Laboratory, where he undertook the first measurements of the mass and speed of alpha particles emitted by radium and polonium.[15] This work seems to have attracted Rutherford's attention to this forty-year-old physicist (Rutherford was then thirty-four).[16]

In early 1907, Rutherford wrote to Mackenzie proposing that he become a fellow of the Royal Society of Canada. He pointed out that while there was currently no vacancy, his departure would leave an opening. In the mean time, he asked him to forward a list of his publications, adding: "I think you will find it interesting to belong to the Society ... For it is the only organized scientific body in the Dominion and will undoubtedly grow in importance with time. At present the meetings of our section ... are not very exciting (between ourselves). But there are great possibilities for the future."[17]

Mackenzie's election gives us a good idea of how the society functioned. Since section III was composed of astronomers, chemists, mathematicians, and physicists, the elections often resulted in negotiations among the members of each discipline. Such dealings must have been quite common, since in his 1898 report the society's secretary suggested that the number of fellows for the scientific sections be increased to thirty and noted: "I may also mention a well known fact that several years past a difficulty has existed in the third section on account of a rivalry between some branches of scientific labour with the result that vacancies cannot be filled up under the rules."[18]

The rules required election by simple majority. This could not be obtained without the support of chemists or mathematicians. Two weeks after having mentioned the possibility of election to Mackenzie, Rutherford wrote:

As I told you in my letter, I was uncertain whether it would be advisable to propose your name this or next year. In the interim between your and my letter, I find that many of the McGill people are committed to another candidate whose name has been up for some time. There are already, I believe, three nominations and only one vacancy, so I think it would be advisable to postpone your nomination till next year ... I shall put your nomination for next year in good hands and hope you will be safely elected.[19]

In fact, there were two nominations. The vote had been divided, and, given four abstentions, nobody was able to garner a majority, and thus no new member was elected in 1907.

Before his departure in May 1907, Rutherford handed over Mackenzie's dossier to his colleague H.T. Barnes, putting him in charge of Mackenzie's election. In February 1908, Barnes wrote Mackenzie, telling him that he had secured "the promise of some of the Toronto people to vote for you and I am very hopeful of the election coming off this year."[20] Although unable to attend the meeting in Ottawa, Mackenzie was finally elected in May 1908.

Despite the frequent negotiations that preceded each election, the number of physicists in section III increased regularly in number and proportion until the mid-1920s. Physicists were closely followed by chemists. Few in number, the mathematicians and astronomers were forced to ally themselves with physicists in order to elect one of their own.[21] Whereas in 1900 Rutherford and MacGregor were the only representatives of the generation of researchers (Cox representing the generation of professors), this number rose to nine in 1915 and to twenty-five in 1929. By proportion, physicists represented 12 per cent of the members of section III in 1900 and 25 per cent in 1915. This proportion reached 30 per cent in 1921 and, until 1939, oscillated between 37 and 40 per cent (Table 4.1).

This growth was paralleled by that of positions in physics in Canadian universities. As we saw in the preceding chapter, these posts were now held by young physicists seeking to create institutional conditions favourable to research. The new members all had research training, and the majority had doctorates. Of the thirty-six physicists elected between 1902 and 1939, there were representatives of all the principal Canadian universities. Six were professors at Toronto and seven at McGill, and, among the twenty-three others, eight were trained at Toronto and two at McGill.

Having become the dominant force in section III during the 1910s, researchers were now in a position to reform the functioning of the Royal Society and its *Transactions*.

BETWEEN ACADEMY AND DISCIPLINARY
ASSOCIATION

The rapid increase in the number of section III members suggests that these people saw the society less as an academy constituted by the national scientific élite and more as an association where members active in the discipline would meet to discuss research. In 1914, 42 per cent of physicists who had a university position were members of the society. This proportion rose to 47 per cent in 1928 and reached 45 per cent in 1939. In other words, during the inter-war years, almost half of the physicists who were at least associate professors in a Canadian university were members of the society. Such a proportion could not qualify the group as a scientific academy, and the body resembled more a disciplinary association.[22] In contrast, only seven of the thirty-six fellows elected between 1900 and 1939 became fellows of the Royal Society of London, one of the highest scientific distinctions of the time.[23]

Canadian physicists' position within the Royal Society of Canada was

Table 4.1

Physicist Fellows Elected to the Royal Society of Canada, 1900–39

Year of Election	Active Members (N)	Physicists (N)	% of Total	Fellows Elected
1900	25	3	12	
1902	29	3	10.3	H.T. Barnes
1903	30	4	13.3	J.C. McLennan
1908	29	4	13.7	A.S. Mackenzie
1909	31	5	16.1	F. Allen
1910	34	6	17.6	A.S. Eve, H.A. Wilson
1911	37	6	16.2	R.K. McClung
1913	37	7	18.9	E.F. Burton
1915	40	9	22.5	A.L. Clark, L.V. King
1916	39	10	25.6	H.L. Bronson
1917	37	11	29.7	J. Satterly
1918	40	12	30	J. Patterson
1919	38	11	28.9	
1921	39	12	30.7	R.W. Boyle
1922	42	14	33.3	J.A. Gray, H.L. Hughes
1923	45	17	37.7	A.N. Shaw, R.K. Young
1924	47	17	36.1	H.F. Dawes
1925	50	18	36	L. Gilchrist
1926	53	20	37.7	D.A. Keys, J.K. Robertson
1927	54	21	38.8	H.A. McTaggart
1928	57	22	38.5	S. Smith
1929	60	24	40	J.S. Foster, G.H. Henderson
1930	62	25	40.3	R.J. Lang
1932	66	26	39.3	E.L. Harrington, J.L. Synge
1934	71	27	38	T. Alty
1935	74	27	36.4	G.M. Shrum
1936	77	28	36.3	A.B. McLay, D.C. Rose
1937	80	29	36.2	W.H. Watson
1938	79	28	35.4	
1939	80	29	36.2	M.F. Crawford, G. Herzberg

Source: *Transactions* of the Royal Society of Canada.

similar to that of fellows of the Amercian Physical Society (APS), who, in 1934, accounted for a third of the membership of that association. Moreover, the majority of the members of the Royal Society participated in APS activities, and A.S. Mackenzie was a founding member in 1899. McLennan, Rutherford, and Barnes, often accompanied by their students, regularly presented their research results at the society's annual meeting in late December, often held conjointly with the American Association for the Advancement of Science. After the First World War, the APS counted among its members those who occupied the recently opened physics positions in Canadian universities such as T.C. Hebb at British Columbia, A.E. Hennings and E.L. Harrington of Saskatchewan, L. Gilchrist of Toronto, A.L. Clark and A.L. Hughes of Queen's, and A.S. Eve, L.V. King, and A.N. Shaw of McGill.

Canadian physicists' participation in the APS was not limited to presentation of research results. Following Rutherford's election to the council of the APS in 1904 for a three-year term, Canadian physicists maintained constant representation on the society's council. J.C. McLennan was elected twice, in 1909 and 1919, and H.T. Barnes held office from 1913 to 1916, his colleague A.S. Eve from 1924 to 1927, and H.L. Bronson from 1928. When the APS took over the publication of the *Physical Review* in 1913, McLennan, along with R.A. Millikan, was a member of the first editorial board, a position occupied subsequently by L.V. King (1919–21). Although Canadians were in 1915 only twenty-five of the APS's 365 members and in 1934 only thirty-five of a total of 2,055, they maintained almost continual representation on the society's council until the end of the 1920s.[24]

Although the APS was never explicitly mentioned as a model, the most active members of the Royal Society of Canada (McLennan, King, and Eve) also participated in APS meetings and could therefore not help but compare the two organizations on the basis of their efficacy in stimulating research. Although the presence of "several visitors" was often mentioned from 1914 on, the average number of members present at meetings of section III (including all disciplines) was around fifteen between 1900 and 1918 and about twenty-five between 1917 and 1927. Even if we add twenty "visitors," the total number was hardly fifty, with a maximum of twenty physicists.[25]

In December 1921, the annual meeting of the American Association for the Advancement of Science (AAAS), held in Toronto, offered Canadian scientists and physicists in particular the opportunity to compare their activity at the Royal Society with that of the APS, which had taken advantage of the occasion to hold its twenty-third annual meeting in Toronto. About 150 physicists attended the different sessions, and seventy-three papers – ten Canadian – were presented. A symposium was held on quantum theory, and McLennan, vice-president of the AAAS, presented a paper on atomic nuclei and electronic configurations.[26]

Several months earlier, in May, the physicist fellows of the Royal Society had met to listen to the presentation of fifty-nine papers, only half of which were in physics, before a public composed of more than fifty astronomers, chemists, mathematicians, and physicists. Confronted with such a dispersion, it is likely that the physicists of section III would have wished for an increase in their number in the society, even if the new arrivals were not capable of exceptional contributions, in order to improve the level of exchange within the section and in order not to be dominated by chemists, whose number had grown as rapidly as the physicists'.

Starting in 1918, the number of papers presented in section III grew rapidly, as can be seen in Figure 4.1. While the average number of papers presented between 1900 and 1915 was nineteen, it oscillated around 104 between 1923 and 1930. In physics (Figure 4.2), the number doubled between 1915 and 1927, the average being thirty-nine during the 1920s, compared to fifteen during the preceding decade. In general, growth was concentrated in two disciplines, chemistry and physics. Since 1917, they had benefited from considerable financial aid from the NRC, in terms of both student scholarships and research grants.

The number of papers having reached 100 in 1923, it became customary to divide the section into two groups. The chemists met among themselves, while the physicists, mathematicians, and astronomers formed a second group, dominated by physicists. More general papers were presented in a plenary session. In May 1925, for example, forty-nine papers were presented by chemists before chemists; forty-five before other members, of which thirty-five dealt with physics; and fourteen to a plenary session. On this occasion, members suggested that the society's council create a new section for chemistry, given the large number of papers presented and the increasing number of chemists, who, despite their worthwhile contributions, could not be elected fellows under the present rules, for they had to compete with physicists, chemists, mathematicians, and astronomers for each vacant place. Such a solution had been adopted in 1918 and had given birth to section V, for biologists. At that time, the rapidly increasing corps of biologists, compared to geologists, had resulted in a disequilibrium in section IV. This time, for reasons unknown, the council rejected the proposition, and chemists and physicists were forced to continue their cohabitation.[27]

The physicists themselves proposed regional meetings, like those of the APS. Following discussion, they realized that, in practice, only scientists in Quebec and Ontario – that is, from Toronto, McGill, and Queen's – would be able to attend the meeting planned for the end of December. As the Royal Society was unable to defray costs, the meeting did not take place.[28]

The increased number of graduate students presenting papers in sections III and IV led members of these sections to discuss integration of young researchers into the society's activities. In the early 1920s, a committee to which McLennan belonged recommended that the maximum number of fellows per section "be enlarged sufficiently to

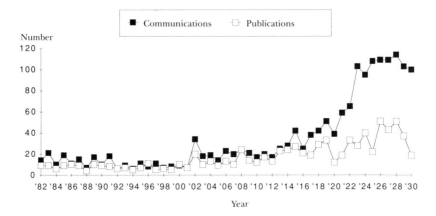

Figure 4.1
Communications and Publications in Section III of the Royal Society of
Canada, 1882–1930

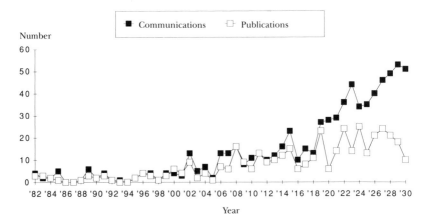

Figure 4.2
Communications and Publications in Physics at the Royal Society of Canada,
1882–1930

permit increase in the number of active workers of recognized ability." The committee also suggested that mechanisms be put into place which would allow ageing and less active fellows to be placed on a retired list, where they would conserve all the honours of belonging to the society while making room for younger researchers.[29]

The problem of young researchers' representation preoccupied many physicists. With regard to the election of new members, King remarked to Patterson: "In the course of the next few years we will have to include in the list such men as D.A. Keys, G.H. Henderson, E.S. Bieler and one or two others of the rising generation. How we shall ultimately manage it is a difficult question under the existing Royal Society constitution."[30]

A biologist and member of section V, Robert Thompson suggested in 1923 that the council consider a Canadian association for the advancement of science, where "the younger scientific men might meet with the older ones, to their mutual advantages, and where businessmen and others interested in science, as well as the public generally, might be brought together for the advancement of the common cause."[31]

Although they opposed creation of such an organization, the different sections did favour "the extension of an invitation to junior scientists to the Royal Society of Canada." According to a proposition, this invitation could be put into practice "by the creation of an Associate Member class which in each section might provide a place for all serious workers in the subjects dealt with by the respective section."[32] Nothing concrete emerged from these discussions, but they none the less indicated the rapid growth in research following the First World War, which gave rise to a young generation of researchers trained in Canadian universities.

The decision of some leading Canadian physicists to work within the Royal Society in order to shape it according to their interests contrasted clearly with American physicists. At the turn of the century, the latter had founded the APS on the pretext that few physicists would be elected to the National Academy of Sciences. They thus decided against redefining the academy in terms of their own interests. Nor did they recognize themselves in section B of the AAAS, which, beginning in the 1880s, was restricted to physicists. As Kevles suggests, the minor influence of the AAAS between annual meetings probably made it inappropriate as an organ for disciplinary representation.[33] Because the Royal Society existed only at its annual meetings, so to speak, Canadian physicists attempted between the two world wars to overcome this obstacle by using the Associate Committee on Physics and Engineering Physics of the NRC, set up in 1919, as an organ of representation (see chapter 6).

If the Royal Society's publication of a special bulletin was a solution for the exceptional case represented by Rutherford in Montreal, it was evident to members of the society that a long-term solution, which responded to the needs of the majority of Canadian producers of science, necessitated major reform of the *Transactions*. After Rutherford's departure in 1907, delays in publication continued to concern the most active members of the scientific sections. As the *Transactions* contained contributions from all sections, complaints were frequent about tardy contributors who neglected to forward corrected proofs. Strict deadlines were suggested. Late contributions would be refused, and articles would be printed on separate pages, so that one author might not impede another. Offprints could also be circulated if the volume had yet to be bound.[34]

There was a financial aspect to the publication problem. Since 1882, the dominion government's annual grant to the society had remained at $5,000, while costs of printing the annual volumes had risen steadily. In 1912, the president of the society, W.D. Le Sueur, approached the government to ask that the grant be increased to $8,000 a year. Pointing out that "the cost of printing ... has largely increased and it is only by reducing the size of the volume that the cost of production is kept within safe limits," he added that "this grant would enable the Society to improve its publications and to meet the growing demand for their gratuitous distribution." Finally, the increase would encourage research, whereas, until now, "it was not found possible to give much, if any, assistance for original scientific research or literary production."[35]

The Royal Society obtained the increase, and, during the May 1914 meeting, members of the scientific sections suggested that part of the grant be used to encourage research by creating a medal to honour a Canadian scientist or literary person. After consulting the sections, the council concluded that "what was greatly needed was a medium of immediate publication for literary and scientific papers." Although the literary sections were included, it is clear that the council's project was intended to assist the scientific sections. The increase in the annual grant allowed the society to consider publishing the *Transactions* four times a year, by bringing sections I and II together in one issue, while each of the three scientific sections would be printed separately. Each section's publication committee would thus become the editorial committee to which members would send their papers for evaluation. As for non-member scientists, they would be able to submit papers, provided they were previously accepted by a member.[36]

Rather than appearing a year late, papers presented at the society's annual meeting would be printed four times a year, in June, September, December, and March, and could include papers not presented before society members. The council noted that these changes should encourage other Canadian scientists to publish their results in Canada. They admitted that, heretofore, most scientists published abroad in order to avoid publication delays. Moreover, for members of the scientific sections "priority would be more easily secured. Under present conditions it is almost impossible to secure such priority." Finally, council members noted that the proposed changes would facilitate circulation of research results and would improve the society's reputation.[37] Given the urgency of these changes and the resultant costs, the society decided not to go ahead with the medal. Scientists had to wait another ten years for minting of the Flavelle Medal, given in honour of a Canadian scientist's work.

So, beginning in 1914, the *Transactions* of the Royal Society of Canada were published four times a year. Unfortunately, the society's financial situation depended on the goodwill of the dominion government, which in May 1917, reduced the society's grant to $4,000, which forced return to the publication of a single volume in 1919. To compensate, the members of section III suggested that printing be speeded up by rejecting all papers received after the first of July following the annual meeting, so that the volume would be ready by the end of the year. They also asked that authors receive fifty reprints free, in order to facilitate distribution.[38]

NRC TO THE RESCUE

Although this procedure reduced costs of production, it was insufficient, and the society was forced to solicit financial aid from the NRC in order to wipe out the deficit created by publication of the 1919 volume of the *Transactions*. Collaboration between the two institutions was quite feasible: many members of the scientific sections of the society sat on the NRC, and the latter felt responsible for diffusion of the results of scientific work that it had helped fund and that were often published in foreign journals.

The first expression of the NRC's interest in Canadian scientific journals went back to December 1918, when it had received a proposition from J.C. Adams, a biologist working for the Canadian government's experimental farm, proposing a Canadian agricultural science journal. Interested in the subject, council members decided to set up a committee to study publication in general and to report to a later meeting.

Composed of A.S. Mackenzie, A.B. Macallum, and R.F. Ruttan (who became president of the Royal Society in 1919), the committee tabled a report three months later and concluded that once it was equipped with national laboratories the NRC "should have as part of its work and scope a "division of publications' and that it might issue three different set of publications," to one of which "would come papers published by men receiving grants from the Council for assisted researches, and similar papers, when approved by the Council, for which there is no natural medium of publication."[39]

As a number of articles were waiting to be published, the committee suggested that "the Council should assist in having such papers printed as soon as possible, and proposes that the Royal Society be asked to publish at least the most important and be reimbursed for the expense to which [it] may be put at by a grant from the funds of the Research Council."[40] The council adopted the report, and the director, Macallum, took it on himself to discuss costs with the treasurer of the society. At the following meeting, held in early May, McLennan, seconded by the Queen's University chemist W.L. Goodwin, proposed that $3,000 be voted to the Royal Society "for the publication of papers submitted by the Council which meets with their approval."[41] Following the society's annual meeting, held a few weeks later, the council approved a list of sixty papers: twenty in physics, ten in chemistry, eight in geology (section v), and twelve in biology (section iv).[42]

The president of the Royal Society, Ruttan, wrote in his annual report that "it would not have been possible to issue the *Transactions* for 1919, had the Advisory Council for Scientific Research not made a special grant of $3,000 to the Royal Society," aid given "to enable the Royal Society to publish in its transactions the more valuable researches in the fields of Biology, Physics, Chemistry and Geology, instead of having these papers scattered in American and English Periodicals."[43]

Compared to preceding years, section III's activities during the 1919 meeting reached a new summit: fifty-one papers were presented, "a large number of which dealt with researches carried out in England and Canada for the British Admiralty, in relation to anti-submarine warfare and kindred subjects."[44]

In 1920 the government's annual subsidy to the Royal Society returned to $8,000 but the NRC's contribution was still required to avoid a deficit. Exacerbating the society's precarious financial situation was the dramatic increase in papers presented at the annual meeting, which could only have led to greater dissatisfaction among scientists, since, as clearly shown above in Figure 4.1, papers published did not keep up with papers presented as costs were too high.

Physicists were particularly worried about this phenomenon, which began affecting them in 1920 (see Figure 4.2, above), and it was quickly put on the agenda of a meeting of the NRC's Associate Committee on Physics and Engineering Physics. As we shall see in chapter 6, this committee, established in December 1919 and led by J.C. McLennan, was the only official organ that allowed Canadian physicists to discuss organizational problems related to research. During the 17 May 1921 meeting (one day before the Royal Society's meeting), A.S. Eve, seconded by John Patterson, proposed that the committee "ask the Royal Society to take up the question of publishing papers of Section III separate from ordinary annual volume and at intervals throughout the year."[45] The following day, during the meeting, section III members returned to the issue, proposing that, "funds permitting," papers presented by members of the scientific sections be published immediately and that authors receive 100 reprints free.[46]

On 20 May, the NRC, meeting in Ottawa, added its weight to this demand.[47] The publications committee, chaired by A.S. Mackenzie, recommended to members of the NRC that a maximum of $3,000 be put at the disposal of the Royal Society subject to the following conditions: 1 / that three-fifths of the amount allocated for publication of the annual volume be reserved for the scientific sections; 2 / "that the papers of Section III, IV, V be printed as soon as they have been handed in ready for publication and have received the approval of the Editorial and the Printing Committee of the Royal Society," and 3 / that "100 reprints of each papers be given gratis to the authors."[48] For the third consecutive year, the NRC had therefore agreed to give the society $3,000 to assure annual publication of the *Transactions.*

The following year, the Associate Committee on Physics and Engineering Physics met in Ottawa on the eve of the Royal Society of Canada's meeting and again discussed publication of articles produced by Canadian researchers. A.S. Eve suggested again that papers judged important be speedily published. In general, members agreed that "a satisfactory solution of the problem would not be reached until there is a single physical journal."[49] The members of the committee directed H.L. Bronson, R.W. Boyle, and L.V. King to write up immediately a report that John Patterson – physicist, secretary of section III, and committee member – could submit that evening to the council of the Royal Society. The one-page report directed the council's attention to "the unsatisfactory state of affairs relating to the publication of scientific work in Canada. As a result of existing conditions, the majority of important papers are submitted to monthly scientific journals in Great Britain and the United States which ensure world-wide distribution to private subscribers, University departmental libraries and research laboratories."[50]

The report noted that the present system of publication was inappropriate in three respects. First, although important articles could be published between annual meetings, there was no particular way of doing this. Second, the authors of the report were opposed to publication of a single volume including papers from all sections, because of the inconvenience this caused in library cataloguing and because of the superfluous costs occasioned in distribution of papers "that are of no special interest to scientific men." Finally, distribution of the *Transactions* was too restricted. "Publications do not reach a sufficiently extensive scientific public, as do monthly journals." The committee therefore suggested to the council that scientific papers be published at regular intervals in a periodical sold at a fixed price and that an editor and publishing directors be named.[51]

On reception of the report, the council decided to strike another committee, chaired by Patterson, to study the situation. Members of section III had already approved the recommendations and had added that their papers should appear separately from the other scientific sections.[52] Patterson's report, prepared for the May 1923 meeting, advanced the same solutions first put into practice in 1914: publish the scientific papers in different issues, twice yearly. Papers could be submitted through the year and judged by two members of the section concerned. Moreover, as members had long demanded, Patterson suggested that authors be given 100 free reprints.

Even though these changes did not transform the *Transactions* into a true scientific journal as members of the NRC's Associate Committee had wished, the council of the Royal Society accepted Patterson's propositions. As might be expected, comment came above all from scientific fellows. In fact, section II was not convinced of the necessity of these changes, even though it claimed to understand the reasons offered by the scientific sections. The latter approved the project, and section IV recommended joint action with universities and the government to help in the publication of scientific papers. This suggestion was taken up one month later by H.M. Tory, a member of section III, president of the University of Alberta, and a member of the NRC, at the National Conference of Canadian Universities held 14–16 June. Following this intervention, most universities agreed to participate in financing the *Transactions.*[53]

While discussion took place on the financing of these transformations, J.A. Gray was concerned with publication of his latest research results. At the beginning of November 1924, he wrote to Patterson to inquire into the possibility of publishing several articles in the *Transactions* within a reasonable period of time. The editor-in-chief replied "Nothing definite was settled in regard the publication of one number in July and one in January."[54] Several months later, Tory, now head of the NRC, summarized the situation before members of the NRC, saying that since "no progress is coming for the moment," the body would have to continue financing the Royal Society "as a temporary measure."[55] During the following meeting, the NRC agreed to furnish $750 to pay the deficit occasioned by publication of the 1924 volume of the *Transactions*.

None of this activity radically modified the functioning of the society's journal, and in December 1925 J.C. McLennan complained of the difficulties he faced in publishing quickly the results of his latest research, on the structure of the spectrum of manganese. The NRC then adopted a resolution asking the Royal Society "to publish immediately the papers by McLennan and Mclay on the spectrums of manganese."[56] At about the same time, university contributions finally allowed the Royal Society to publish two issues of the scientific sections "and thus carry out the wishes of the Society when their method of publication was revised."[57] However, difficulties arose the following year, and publication of the second issue was delayed.[58]

In the early 1920s, members of the NRC, who formed the Canadian scientific élite, believed that reforming the *Transactions* of the Royal Society would respond to the needs of Canadian scientists in terms of the diffusion of research results. The slowness of the changes gradually led them to modify their position.

CHAPTER FIVE

The Birth of the Canadian Journal of Research

While John Patterson was presenting his report to the Royal Society of Canada in May 1923, members of the NRC were discussing the financial aid accorded the society annually since 1919 for publication of scientific papers and had concluded that "the desired objective could be better obtained by some other method, and that no further grants should be made to the Royal Society of Canada for this purpose."[1] The members of the Associate Committee on Physics and Engineering Physics did not agree. They believed that "the Transactions of the Royal Society of Canada is the only established medium for general scientific papers in Canada and is undoubtedly the most suitable place for making the work of the Honorary Advisory Council known to Canadians and the world."

The committee optimistically recalled that the society's publication methods had been modified and improved to the extent that "the prompt appearance in creditable form of the results of research work is assured." Finally, the committee felt that publication costs would be even higher if the NRC took charge of publication itself. It therefore recommended that the NRC consider "the advisibility of continuing the publication of the research work of the Council in the *Transactions* of the Royal Society of Canada by making a reasonable and proportionate appropriation for that purpose."[2]

The NRC thus continued to finance the society's deficit on an annual basis, but the new director, H.M. Tory, became more and more convinced that this aid would not solve the problem in the long term. In February 1925, he turned to Walter Murray, president of the University of Saskatchewan and a member of the NRC, to express his opinion of the Royal Society's repeated requests for financial aid from the NRC and Canadian universities:

With regard to the Royal Society's request, I am not personally in favour of tying up the university to anything in the way of permanent assistance to the Royal

Society of Canada and in this I am sure I am in agreement with most of the younger scientific men of the country. I have been hoping that we could get far enough forward with the research organization in Canada that a nucleus of publication work could be begun. We shall never get recognition scientifically until we have some scientific journals of our own.[3]

For this inveterate founder of institutions, creation of a scientific journal was supposed to be the normal outcome of greater numbers of researchers and higher productivity. As the NRC had played a leading role in the development of the Canadian research system, it had to put into place the last important piece of the system – by creating the *Canadian Journal of Research* (*CJR*).

As we have seen, discussions concerning publications followed the same rythm as the annual meetings of the Royal Society, which were held the third week of May. The NRC's decision to take charge of the creation of a new scientific journal, finally responding to the needs of Canadian scientists, occurred therefore just after the May 1926 meeting of the Royal Society.[4] A committee, composed of a majority of physicists,[5] was struck to make plans for the new journal and to report to council members.

While plans were being made for the new journal, the NRC continued to underwrite the costs of the *Transactions* of the Royal Society of Canada. Between 1919 and 1928, the NRC had contributed close to $17,000 to aid in the publication of papers written by members of the scientific sections of the Royal Society. From an annual value of $3,000 between 1919 and 1921, the NRC's grant fell to $2,000 in 1922 and remained around $1,000 a year between 1923 and 1927.[6]

In July 1929, the committee had completed its mandate and deposited the first issue of the *CJR* (dated May 1929) before members of the council.[7] Conscious of the conflict that publication of the journal might cause between the NRC and the Royal Society, the president of the NRC used the appearance of the first issue to point out to members that he had received:

Several letters of approval of the action of the Council in deciding to establish a Canadian Journal of Research, including letters from the Editors of "Canadian Textile journal' ... There had been no criticism from any source of the decision of the Council to establish such a journal in Canada, but, on the other hand, the action of the Council had been widely commended. The single objection was with regard to the title of the journal, and this objection had been raised only in Toronto.[8]

In a foreword to the first issue, Tory recalled that papers produced as a result of the activities of the NRC, universities, and other research organizations were now sufficiently numerous to create difficulties in terms of rapid publication. Remaining silent as to the role played by the *Transactions* of the Royal Society, he added that these difficulties were "largely due to the fact that there has been in Canada no national periodical devoted to research".[9] Reiterating the same theme in his annual report to the government, Tory specified that "this situation not only resulted in delaying the publication of Canadian scientific papers, but also resulted in loss of credit to Canada for the research work covered by papers published in foreign journals."[10] This last argument, which would have been out of place in a journal destined for scientists who prized foreign publication, was pertinent in a report seeking to convince the government that the NRC's decisions were made in the interests of Canada.

CANADIANS IN PHYSICS:
PERCEPTIONS AND REALITIES

In quantitative terms, Tory's position was well-founded. Between 1900 and 1928, McGill physicists had published 45 per cent of their papers in England, 26 per cent in Canada, and 4 per cent in Germany. At Toronto, where McLennan and his team produced the majority of papers, the distribution was different: 46 per cent were published in Canada, 39 per cent in England, 10 per cent in the United States, and 2 per cent in Germany (Table A7). Whether or not Canada lost credibility as a result of such practices is less evident, for the interpretation of foreign publication is a rather more delicate issue. For some, foreign publication is a proof of high quality, while for others it is a sign of dependence and lack of autonomy.[11]

In fact, spokesmen for scientists called on both interpretations. Addressing politicians, Tory and Ruttan deplored the publication of Canadian work abroad. Yet they both knew that international recognition was to be had only through publication in British journals. It was therefore a question of attaining a certain equilibrium: Canada should have its own high-quality scientific journals, but scientists should not deprive themselves of foreign publication. One thing was certain: Canadians sent what they considered their most important papers to the three main British journals: *Transactions of the Royal Society of London, Philosophical Magazine,* and *Nature.* The fact that almost all papers published in Britain by Canadian physicists between 1900 and 1928 appeared in these three journals shows clearly that Canadians formed a part of the international scientific community. Moreover, these publications represent 41 per cent of the total number of articles published by Canadian physicists during this period: they must have been of excellent quality according to the criteria of the times.

In comparison, between 1894 and 1915, the most productive American physicists published 34 per cent of their papers abroad – for the most part in Britain (56 per cent) and Germany (40 per cent) – while Canadian physicists did the same with 75 per cent of the papers they published between 1900 and 1917 (43 per cent in Britain, 26 per cent in the United States and 5 per cent in Germany).[12] *Philosophical Magazine* published 49 per cent of American papers published abroad, compared to 35 per cent of Canadian papers. *Nature* published 2 per cent of American papers compared to 11 per cent of Canadian papers. Few American physicists published in the *Transactions of the Royal Society of London*, whereas 8 per cent of Canadian papers appeared in that journal between 1900 and 1917. Here we see the effect of the links forged since the 1890s between British and Canadian physicists and the presence of Canadian fellows within the Royal Society of London. Between 1918 and 1929, the proportion of Canadian papers published abroad fell to 54 per cent. A third of these were published in the United States and two-thirds in Britain; 25 per cent appeared in *Philosophical Magazine*, 14 per cent in *Nature*, 11 per cent in *Physical Review*, 26 per cent in other American journals, and 5 per cent in other British journals.

Unlike their Australian colleagues, who published almost all their papers in British journals, and their American colleagues, who, like the Germans and the British, published mostly within their own frontiers, Canadian physicists had a more diversified publishing practice – one in fact comparable to that of Indian physicists, who published 60 per cent of their foreign publications in Britain, 24 per cent in Germany, and 14 per cent in the United States (Table 5.1).

To evaluate the visibility of Canadian publications in the international field of physics, I have assembled Table 5.2, which shows the number of citations received between 1920 and 1929 by article cited and published in Canada and abroad. In the United States, Canadian publications achieved a visiblity proportionate to the number of articles published in American journals: 26 per cent of the total number of citations received for 23 per cent of the total number of articles published in this country. The situation was similar in Britain: Canadians published 38 per cent of their papers in that country and received 32 per cent of their citations from British journals. In Germany, however, they were much more visible: Canadians received 33 per cent of their citations from German journals, whereas, between 1900 and 1928, they published only 2.5 per cent of their work in these journals.

More interesting, the distribution of citations according to country is more or less the same for articles appearing in Canadian journals as for those appearing in foreign journals, although articles published in Canada were slightly more visible in American journals and slightly less so in German ones. On average, each cited article received 3.1 citations

Table 5.1

Place of Publication by Physicists from Different Countries, 1920–29

Country of Origin	Place of Publication (% of Total)		
	United States	Great Britain	Germany
Argentina P = 61; A = 12	0	2	93
Australia P = 66; A = 33	0	100	0
Canada P = 193; A = 73	42	57	0
Germany P = 2,715; A = 2,439	0.4	1	98
Great Britain P = 1,620; A = 1,516	1	83	1
India P = 268; A = 71	14	62	24
United States P = 2,867; A = 2,769	89	7	3

Source: Henry Small, A Citation Index for Physics, Report to the National Science Foundation, Sept. 1980.
Note: P = papers; A = authors.

from foreign researchers – significantly higher than the international average, 2.0.[13] Moreover, the mean number of citations accorded to articles cited that were published in Canada (4.1, without counting citations attributed by Canadians) was comparable to that accorded articles cited that were published abroad (3.0). These figures suggest that the quality of work was the same in the two cases and that, as long as the work was known to foreign physicists, the articles published in Canada obtained the same citation rate as those published abroad. However, as only 20 per cent of the total number of citations received by Canadian physicists were for articles published in Canada, while between 1900 and 1929 Canadians published 36 per cent of their papers in Canadian journals, work published abroad clearly had a better chance of being cited. Although Canadians were in a better position than their colleagues in other British colonies and dominions, these two phenomena demonstrate the importance Canadian scientists accorded the adequate distribution of their work.[14]

Table 5.2
Distribution of Citations to Canadian Physics Publications by Country of Origin,
1920–29

Country of Origin of Citation	Place of Publication of Canadian Papers					
	Canada		*Outside Canada*		*Total*	
	N	%	N	%	N	%
France	14	5	66	7	80	6
Germany	74	30	340	34	414	33
Great Britain	81	33	317	32	398	32
United States	79	32	241	24	320	26
Other Countries	0	0	26	3	26	2
Total	248	100	990	100	1,238	99

Source: As Table 5.1.

The visibility of the *Transactions* of the Royal Society of Canada is also indicated by the *Physics Citation Index*, which shows that for the period 1920–9 the *Transactions* received about the same number of citations as the three main Indian journals combined. The respective visibility of the Canadian and Indian journals in different countries reflects physicists' foreign publication practices: the Canadian journal and the Indian journals had approximately the same visibility in Britain, while the latter were more visible in Germany. The *Transactions* were more visible in the United States (Table 5.3). The Australian journals, in contrast, obtained less than twenty citations during the period, which suggests that Australian physicists were less well organized than their Indian and Canadian colleagues.

MANAGING THE NEW JOURNAL

If creation of the *Canadian Journal of Research* (*CJR*) responded to a need long felt by Canadian researchers, its adequate distribution still had to be ensured. At the NRC, this task devolved to the Division of Research Information, which was in charge of all NRC publications. Particular attention was paid to diffusion of the journal which soon circulated more widely than the Royal Society's *Transactions*. In 1925, for example, only 250 copies of the section III *Transactions* were distributed. For economic reasons, this was reduced to 150 in 1928. Aside from copies sent to members and Canadian libraries, it is doubtful that more than 200 copies of the *Transactions* circulated abroad. There was no committee responsible for its distribution; fellows often complained of the

Table 5.3
Citations to Canadian and Indian Journals, 1900–29

Country of Origin of Citations	Canada*		India†	
	N	%	N	%
France	9	5	22	11
Germany	30	18	63	31.7
Great Britain	77	45	80	40.2
United States	55	32	31	15.6
Total	171	100	196	98.5

*Transactions of the Royal Society of Canada.
†Proceedings of the Indian Academy of Science; Indian Association of Culture and Science; Indian Journal of Physics.

uselessness of the *Transactions* for disseminating their research results and demanded that each author receive at least 100 reprints.

Eight months after publication of the first issue, the *CJR* already had 477 subscribers, to which were added forty-eight institutions that exchanged their journals with the NRC.[15] In 1933, the *CJR* was distributed in twenty-three countries. To make the journal more widely known and especially to increase subscription, the division also produced a list of the titles and summaries of the papers published and distributed the list to scientists whose names were often furnished by authors.[16] In December 1945, 833 copies of section A, printed separately since 1944, were distributed: 242 in Canada, 389 in the United States, 98 in Britain, and 99 in other countries. It is likely that the *CJR* reached the majority of its potential clients. Section B of the *CJR*, devoted to chemistry, had 1,100 subscribers, distributed in much the same way as section A.[17]

Conceived "for the publication of results of work carried out under the auspices of the Council," the *CJR* was also "open for suitable papers from any Canadian research workers."[18] As Figure 5.1 indicates, creation of the *CJR* led to a sharp fall in the number of physics papers published by the *Transactions*. The same was true for chemistry papers and, to a lesser extent, biology. Toward the end of the 1930s, the *Transactions* published mainly the presidential addresses. In section III, there were approximately five papers per volume, furnished mainly by mathematicians, who did not have their own journal until 1945.[19]

After having published two bi-monthly issues, the *CJR* went monthly. Each issue contained mainly contributions from chemists, physicists, and biologists (Figure 5.2). Starting in July 1935, two different issues were

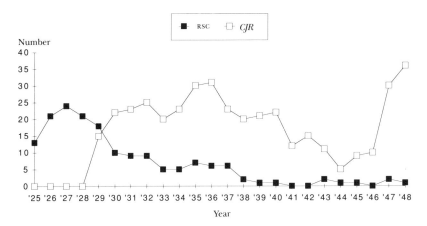

Figure 5.1
Physics Papers in RSC *Transactions* and in *CJR*, 1925–48

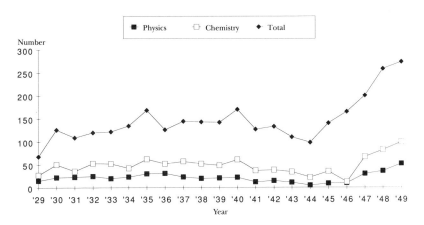

Figure 5.2
Papers Published in *CJR* by Discipline, 1929–49

printed: one contained articles in botany (section C) and zoology (D), and the other, articles in physics (A) and chemistry (B).

The papers came mainly from university researchers. Members of the NRC labs contributed approximately 20 per cent of the total. Until 1932, the majority of authors benefited from the NRC's financial aid. However, the opening of laboratories in 1932 and the economic crisis sharply reduced the amounts accorded individual researchers, which explains the rapid decline in the number of papers produced with the help of grants (Figure 5.3).[20]

Although the title of the journal was in English, French articles were also accepted. As has already been mentioned, in francophone universities research began in chemistry, so that it is not surprising to find that the first articles published by francophone researchers in the NRC's journal were the work of Joseph Risi, professor at the School of Chemistry at Laval University, who in 1935 published two articles in collaboration with his students. The first bore on the aromatic nature of maple products, while the second, published in two parts, was a contribution to the study of polymerization. The maple-water research project resulted in a number of other publications signed by Elphège Bois and Eustache Nadeau, also of the School of Chemistry. Chemists at the Université de Montréal began publishing in the *CJR* only in 1940. At that time, Léon Lortie, professor in the chemistry department, and his student Pierre Demers, published two studies on the physicochemical properties of alkaline carbonates. Having subsequently switched to physics, Demers published the first physics article in French in the *CJR*, in March 1944. The article was devoted to Maxwell's Demon and the second law of thermodynamics. Demers wrote the only three French-language papers that appeared in the journal before the end of the war. In all, only thirty-one papers were published in French between 1935 and 1945: twenty-one in chemistry, five in botany (three of them by Pierre Dansereau), and two in biology (by A.R. Gobeil). This represented only 2 per cent of the total for the period. Although the number of francophone researchers grew rapidly after the Second World War, the title page of the *CJR* did not become bilingual until January 1973.

One should not expect that creation of the *CJR* instantaneously increased the proportion of papers published in Canada by Canadian physicists. In fact, during the decade preceding the Second World War, Toronto physicists continued to publish about half (42 per cent) of their papers in Canada, 33 per cent in England, 21 per cent in the United States, and 3 per cent in Germany. At McGill, where physicists' productivity was three times less than it was at Toronto, 38 per cent of the papers continued to be published in England, especially in the *Transactions of the Royal Society of London* and the *Philosophical Magazine*,

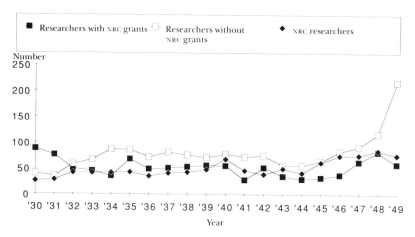

Figure 5.3
Origin of Papers Published in *CJR*, 1930–49

20 per cent in the United States, 28 per cent of which appeared in *Physical Review*, and more than a third (48 per cent) in Canada.[21] NRC researchers published all their results in their journal (Table A7).

In essence, the *CJR* replaced the *Transactions* of the Royal Society of Canada and ensured improved diffusion of research results produced in Canada. However, the journal did not attract a greater proportion of Canadian physicists' production. These scientists seem to have found a satisfying equilibrium between national and international publication.

TOWARD DISCIPLINARY CONTROL
OF THE *CJR*

Although it served the members of the Canadian scientific community, the *CJR* was not controlled by them. To somebody with the habit of perusing scientific journals and looking at the editorial committee and the rules governing contributions, the *CJR* must have appeared rather curious and might easily have been identified as an in-house publication, which it really was not. A scientist wishing to publish in the journal would have found only on its back cover the following inscription: "Address all correspondence to the National Research Council, Ottawa, Canada." Publication of the journal was entirely in the hands of the adjunct director of the division, the chemist Walter W. Thompson, assisted in the beginning by the physicist Dorothy DesBarres and, starting in 1940, by an adjunct director, Pauline Snure, who took charge of production. As Norman T. Gridgeman recalled: "Stemming from convenience and economy, the editing and refereeing had become institutionalized as in-Council business: all manuscripts came to the editorial office on

Sussex Drive, and then they were sent along the corridor to whichever resident scientist was most expert on the subject. He would then pass judgement, taking the task as part of his job. It is not necessary to spell out what many non-NRCC scientists thought of this cosy setup."[22]

Although researchers naturally had enough free time to criticize individually the functioning of the *CJR*, most did not have official organs of representation which could have served as spokesmen and negotiated "in the name of the discipline." During the 1920s, physicists had learned to use the NRC's Associate Committee on Physics and Engineering Physics as a means of representing themselves to the Royal Society of Canada. However, this committee, which had always been chaired by McLennan, had not met since 1930 and disappeared definitively in 1932. Moreover, the journal was not truly disciplinary – a single issue contained papers in biology, chemistry, physics, and zoology – which allowed no single discipline to control it and to elaborate policy in terms of its needs.

Establishment of an editorial committee composed of representatives of the different scientific disciplines covered by the journal only followed its division into disciplinary sections and negotiations with the Royal Society, which until 1945 remained the most legitimate representative of the interests of Canadian scientists.

Even chemists, who had formed a representative organization in the 1920s, did not until 1945 demand the right to oversee publication of chemistry articles in the *CJR*. It is likely that the existence of several organizations, such as the Canadian Institute of Chemistry, founded in 1921, and the Canadian Chemical Association, founded in 1928, kept chemists apart and retarded their taking charge of the chemistry section of the *CJR*. It was only after the fusion of these organizations in 1944 that chemists addressed the NRC. United within the Chemical Institute of Canada, they negotiated their representation on the editorial committee of the *CJR*, trading off this representation against recognition of sections B (chemistry) and F (technology) as "official research organs" of the institute.[23]

Probably conscious of the fragile position of their journal, NRC members discussed publication of the *CJR* during the May 1938 meeting and decided "to again approach the Royal Society of Canada with a view to closer co-operation between the Council and the Royal Society in the publication of scientific papers in Canada."[24] With its usual slowness, the committee system of the Royal Society was set in motion. Following a meeting between the presidents of the two organizations, a report was presented to the scientific sections of the Royal Society. The members of section III then formed a committee, which in May of 1940 concluded:

It is seen as essential that to be successful a journal, or section of a journal, must have an editorial board consisting of outstanding investigators in the fields in question, which shall have the responsibility of determining policy and maintaining standards. In accordance with a principle already accepted by the two organizations, the Society might well nominate the personnel of such editorial boards, who would be appointed by, and responsible to, the Research Council.[25]

Section III members endorsed, in principle, the fusion of the Royal Society's *Transactions* and the *CJR* but put off a final decision until the next meeting. In May 1941, the Royal Society's committee finally decided not to fuse the two journals, proposing rather collaboration with the NRC through an editorial committee composed of members of Sections III and IV and representatives of the NRC.[26] The report was accepted by members of the sections concerned, and the joint editorial committee came into existence in 1942. It was composed of eight members (four named by the Royal Society and four by the NRC), to which was added an editor-in-chief, named by the NRC.

Each section was directed by an editor, named for three years and chosen from among members of the editorial committee. For section A, which covered physics, the first editor was J.K. Robertson, of Queen's University: A.N. Shaw, of McGill, succeeded him in 1947. The members named by the NRC usually included a physicist. Thus J.A. Gray, of Queen's, was a member of the editorial committee until January 1948, when he was replaced by G.H. Henderson, of Dalhousie, who, two years later ceded his position to his colleague J.H.L. Johnstone.

The arrival of scientists on the editorial committee led to greater autonomy for each journal section. Starting in January 1944, each section was published separately, even though the first issues were rather slim, sometimes containing only one article. The physicists, as well as the chemists and biologists, finally had their disciplinary journal. However, it was necessary to wait several years more for each section of the *CJR* to acquire its own name.

EPILOGUE

In January 1951, the *CJR* disappeared and was replaced by six different journals including the *Canadian Journal of Physics* and the *Canadian Journal of Chemistry*. As we have seen, in giving themselves a representative organization, chemists had been able in 1945 to address the NRC directly in order to claim a position on the editorial committee of the journal, alongside representatives of the Royal Society and the NRC. That same year, physicists meeting in Toronto founded, for reasons analysed in chapter 7, the Canadian Association of Physicists (CAP). The logic of representation then came into play for physicists, and, in 1951, it was their turn to demand from the NRC, in the name of Canadian physicists, participation in the administration of the newly founded *Canadian Journal of Physics.*[27] As the result of an agreement reached with the NRC, members of CAP were able to subscribe to the journal at a reduced price. On the one hand, the agreement gave the NRC a means of making the journal more profitable; on the other hand, it transformed the *Canadian Journal of Physics* into a "national journal" with which Canadian physicists could identify.

This change, in the early 1950s, marked the end of a long process – from the emergence of a new research-oriented practice of physics at the end of the nineteenth century, to its institutionalization in the 1920s, to the birth of CAP. An official organization, charged with the defence of the interests of Canadian physicists, CAP also sought to diffuse and impose an image of physicists which gave this category of social agents a recognized social existence.

From a historical point of view, establishment of such a representative organization was inseparable from all the struggles of scientists to institutionalize their practice of physics. The next three chapters follow Canadian physicists as they develop means of representation from the middle of the First World War on.

PART THREE

Changing Definitions

CHAPTER SIX

In Search of a Collective Voice

As we have already seen, the "physicist" as we today know and recognize him or her, has not always existed. In order to have a complete social existence, the "physicist" had to be "born" twice. First, there had to be agents whose practice was based more on research than on teaching and whose emergence I analysed in chapter 1. However, it was also necessary, and this is the second "birth," that this practice of physics define a group recognized as socially distinct from other categories of social agents. In other words, in order to be thought of socially, physicists had to become a social category.

In Britain, for example, physicists realized during the First World War that few people and few governmental organizations knew that physicists existed. In fact, the category "physicist" simply did not exist in government registers, and consequently physicists, unlike chemists, could not be recognized and had no status.[1] This new consciousness led physicists to ask themselves how they could improve "the professional status of the physicist" as one of them put it.

DEFINITIONS

Generally speaking, historians and sociologists of science interpret this status claim as a sign of the "professionalization" of scientists. A large number of studies describe the formation of a scientific discipline in terms of "professionalization." This term is rarely defined: authors who use it often confuse the different ways in which actors use it in their discourse. As we shall see, the physicists who between roughly 1945 and 1950 wished to construct a physics discipline in Canada opposed another group, which hoped to form a profession somewhat like that of the engineers. To analyse this process simply in terms of professionalization would not allow one to distinguish these two groups who sought to construct different identities for the physicist.[2]

Following Terence Johnson and M. Sarfatti-Larson, I believe that use of the term "professionalization" should be limited to describing the process of the control of an occupation, with the profession understood as a particular mode of control.[3] This distinction is necessary in order to recognize the different uses agents have for this term. For example, as we shall see, university physicists use the term "professional" in a rather ethical sense, whereas industrial physicists – seeking to imitate engineers, doctors, and lawyers, who have succeeded in controlling access to their occupations – have a legal understanding of the term. These two uses are, moreover, the result of agents circulating in different markets. While the discipline is a closed market – where producers of knowledge are also consumers, so that scientists' potential clients are also their "rivals" – the profession is characterized by agents circulating in an open market, where the clients are not themselves producers of knowledge.[4]

By not distinguishing these two uses of the term "professional," which correspond to two practices of physics – one confined to the university and the other open to the industrial milieu, where physicists are in competition with engineers, who tend to monopolize this type of occupation – historians often confuse two distinct processes. One consists in the construction of a scientific discipline, with its associations, meetings, journals, medals, and official representatives; the other seeks to circumscribe a profession by monopolizing access to the titles and posts associated with it.[5]

The rather vague use of the concept of professionalization also takes for granted the existence of scientists as agents without distinguishing individual and collective levels of existence. To return to the British example, the physicist claiming professional status in 1919 did not doubt his own existence as an agent or a physicist. In his language, it was therefore normal that he wished to raise his status. However, from a sociological point of view, this same physicist did not yet socially exist. What we are given to see is rather the process of the constitution of a social identity or category, which defines the social existence of a group and permits it to be thought of by other institutions that make up society. Because this identity may take many forms, it is necessary to distinguish the general process of the constitution of identity from its particular form, which is the result of a historical process of social construction.

After having described the emergence in Canada of a practice of physics which defines the physicist as an individual agent, and after having shown how these new agents imposed their practice of physics within the university institution and the Royal Society of Canada, it remains to be considered how they succeeded in imposing the social figure of the physicist, the persistence of which is assured by organizations of representation. We shall see how the two world wars provoked rapid development of the physicists' collective identity. The need for organizing scientific work in the First World War led (as we saw earlier) to formation of the NRC, and physicists, after looking at British models, eventually chose a disciplinary format in an associate committee of the NRC. The growing number of "industrial physicists" and federal regulation of collective bargaining during the Second World War led to a five-year internal battle in the late 1940s.

The newly formed Canadian Association of Professional Physicists (CAPP) eventually gave way to the renamed, disciplinary group, the Canadian Association of Physicists (CAP). CAP then sought to respond to the "Big Science" of the 1950s.

EMERGENCE OF AN ESPRIT DE CORPS

In the United Kingdom, the circumstances surrounding the First World War led physicists to become concerned with their social status, or, rather, in our terms, to become conscious of their social non-existence and to seek to remedy the situation by giving themselves a "representative" organization, the Institute of Physics. The situation was different in Canada where, at the time, only twenty physicists had institutional positions, the majority within universities. These researchers were preoccupied with consolidating their positions through imposition of a new discourse, in which research was to be an integral part of the functions of the university. In this context, Canadian physicists' organizational preoccupations were the result of the establishment in 1916 of the National Research Council and its organizational logic.

One of the primary functions of the NRC was, in effect, the organization of scientific and industrial research in Canada. To accomplish this goal, the NRC created associate committees, composed of volunteer scientists, to study particular problems or even to localize problems to be studied by the NRC. In general, if the sheer number of Canadian physicists did not spontaneously lead to the production of a discourse on representation, the logic of the committees did generate the question: "Should we form a Canadian physics society?"

After having put into place a bursary system and research grants, members of the NRC asked themselves what were the problems in each discipline which were of interest to Canada from a national standpoint. During the council meeting of 25 May 1918, A.S. Mackenzie was given the responsibility of contacting the directors of the physics departments of Canadian universities and asking them "whether it would not be advisable to hold an annual conference of the physicists and electrical engineers and perhaps other related scientific men for counsel and advice on general physical problems of interest to Canada from a national standpoint."

Mackenzie asked the directors also to express their opinion "as to the advisability of considering the formation of a Canadian Physical Society."[6] This latter point had not been part of Mackenzie's original mandate, and we can only speculate as to what led him to ask such a question at this time. It is likely that creation of the British Institute of Physics inspired the initiative. Further, many Canadian physicists, such as J.C. McLennan, A.S. Eve, and Robert W. Boyle, were then working for the Board of Invention and Research wherein had emerged, in November 1917, the idea for the Institute of Physics.[7] Moreover, as we shall see, as soon as he returned to Canada in September 1919, McLennan proposed establishment of a Canadian section of the Institute of Physics.

Whatever its origin, the idea of a Canadian physicists' association had been launched. The possibility of advancing such an idea was based on the very existence of the NRC, whose task was to raise questions pertaining to the organization of scientific activity. Without such an organization, the idea might have germinated in somebody's head without having much of a chance to develop, especially in 1918.

In the summer of 1918, Mackenzie received comments from nine physicists, all university professors, who took advantage of the occasion offered by the NRC to express their opinion and to suggest kinds of organization which, until then, they had only been able to discuss informally.[8] Thus Louis V. King reported: "Western Scientists want a more representative corporation than the Royal Society." A fellow of the learned society, King believed that section III "should do for a Canadian physical Society by making subdivisions into physics, chemistry and mathematics."[9] However, as we saw in the preceding chapter, these sections never saw the light of day.

Almost all the physicists mentioned that the limited number of researchers in physics and the great distances separating them inhibited formation of a physicists' association. In summary, Mackenzie wrote in his report, "the time is not ripe for the formation of a Canadian Physical Society." However, many believed that the NRC should form an associate committee of physics and engineering physics. J.L. Hogg, of the University of Saskatchewan, wrote that "it is desirable that physicists be organised so as to get corporate opinion on national physical problems. Form them into an Associate Committee of the Council."[10] As Canadian physicists were apparently unable to organize themselves, it was up to the NRC to provide them with a framework for action.

SPEAKING FOR THE DISCIPLINE:
THE NRC'S ASSOCIATE COMMITTEE

At the 2 December 1919 meeting of the NRC, the Associate Committee on Physics and Engineering Physics was set up, with J.C. McLennan as president. After two years in England as scientific advisor to the Royal

Navy, McLennan had just returned to Canada.[11] The official mandate of the committee was rather wide. It was to act "as a consultative Committee for the Council on problems presented for consideration and solution in these two departments of science, and also to bring to the attention of the Council all matters relating to physics and engineering physics, upon which, in the judgement of the Committee, the Council should take action."[12] Composed of fifteen members, the committee contained ten university professors, including eight physicists, and only two industry representatives. If we add John Patterson of the Meteorological Office of Canada, then there were nine physicists on the committee. It was only nine years later that membership was raised to twenty-two, to give greater industrial representation.

During its effective existence, 1919–31, the Associate Committee held thirteen meetings, in the course of which a wide range of problems were discussed.[13] However, the real importance of the committee lay in the less visible part of its work. Through its various activities, the committee undertook the work of constituting the physicists as a group: gathered together for the first time around a table and called on to think in terms of the NRC, that is, in terms of organization, physicists could not help but discuss their own organization as a group, thus transforming "personal opinions" into the "official opinion" of the discipline of physics, to whose social existence they contributed.

During the second meeting, in mid-January 1921, the committee discussed creation of a Canadian branch of the British Institute of Physics. McLennan had been contacted by the directors of the institute in this regard, and the members of the committee were in favour of the idea of creating an association for Canadian physicists. However, what kind of organization should be set up? Should physicists associate with the Engineering Institute of Canada, create an autonomous organization, or open a branch of the Institute of Physics? The committee's meeting coincided with a physics conference held at the University of Toronto's department of physics, and a meeting was held with physicists present at the conference to discuss the different possibilities.[14]

The simultaneity of these two events was not fortuitous. As president of the Associate Committee and director of the physics department and conference organizer, McLennan was killing two birds with one stone: he assured the success of his colloquium by having the NRC defray transportation costs for physicists, who, once in Toronto, could be mobilized to discuss the creation of a disciplinary organization.[15] Although we do not know the exact number of participants, the meeting took place Monday 17 January at 10 a.m. McLennan was the president, and J. Patterson, the secretary, read the following resolution, submitted by the Associate Committee: "To consider the proposal to establish an Institute of Physics of Canada with Honorary Fellows, Fellows, Associate

Fellows, and Members, and to have the Institute affiliate with some other society or societies such as the Royal Society, the Royal Canadian Institute, the Engineering Institute of Canada, or the Institute of Physics of Great Britain."[16]

Nobody seems to have called into question the legitimacy of those who organized the meeting. Presided over by John C. McLennan, the best-known and most influential physicist in Canada, and patronized by the NRC, a governmental organization, the meeting had all the trappings of legitimacy. It did not run the risk of being perceived as the initiative of a self-interested faction, as happened with those wishing to create an association of professional physicists in 1945.

The "Assembly" decided to form an "Organizing Committee" "to make all necessary arrangements and to report to a meeting in Ottawa in May 1921."[17] It appears that this new committee was little more than a sub-group of the original organizers, the NRC's Associate Committee. The organizing committee included, in addition to McLennan, the ubiquitous president, A.S. Eve from McGill, T.C. Hebb from British Columbia, H.L. Bronson from Dalhousie, A.G. McGougan from Saskatchewan, and J. Patterson from the Meteorological Office of Canada. Before the meeting adjourned, a series of motions was passed, one of which thanked the president and board of governors of the University of Toronto for "the service the University has rendered *to the Country* by holding a conference on Physics."[18] Although an expression of gratitude, the motion also served to identify the national interest with the organizational interests of Canadian physicists.[19]

By meeting in assemblies and committees and by discussing resolutions, Canadian physicists gave themselves the opportunity to perceive themselves as a group with distinct interests. The change came about in much the same way as political demonstrations "transform collections of individuals with common social properties which sometimes are not recognized as such, into self-interested groups which can perceive themselves as a group and be counted as such."[20] The January meeting thus contributed to the emerging social existence of "Canadian physicists" by offering members a group reference other than the one associated with the traditional categories of "professors," "engineers," "American physicists," or "British physicists."[21]

Having originally been named to report on questions of physics and engineering physics vital to the national interest, the members of the Associate Committee now found themselves in charge of the establishment of an association of physicists. Unlike British physicists, who were concerned mainly with problems of social status and salary, members of the committee seemed preoccupied with the esprit de corps of Canadian physicists. Thus, during the May 1921 meeting, Patterson suggested publication of a journal that would circulate among "physicists of Canada." Boyle agreed, suggesting that the journal would have a "unifying influence."[22] McLennan proposed that it would be useful to "get out a leaflet, indicating what is being done in research in physical laboratories." Since the committee had met at the same time as the Royal Society, an attempt was made to hold a general assembly. However, for reasons unknown, "the subject was allowed to remain in abeyance."[23]

Meeting again a year later, in Ottawa at the annual meeting of the Royal Society, the committee decided unanimously that the time had not yet come for the formation of a Canadian physicists' association or even for a Canadian branch of the Institute of Physics.[24] This apparent lack of interest can be explained by the existing institutional arrangements. In 1921, there were approximately thirty physics professors in the Canadian university system, and they had trained few graduate students. Twelve physicists were members of the Royal Society of Canada, and the majority of these were also members of either the NRC or its Associate Committee. The NRC and the Royal Society thus served the physicists' most immediate needs.[25] As we saw in the previous chapter, Canadian physicists had learned to use these two organizations to promote research in physics and to diffuse research results within the scientific community. There is thus no need to invoke lack of a "critical mass" to explain the non-existence of a Canadian association of physicists. Existing institutions, despite appearances, carried out the functions normally attributed to disciplinary associations.

The Associate Committee acted as a sort of executive council of a disciplinary association by focusing on issues essential to reproduction of the discipline, such as the quality of science teaching in high schools and the role of the physicist in industry. When, for example, at the first meeting of the committee, the question of physics teaching was raised, McLennan remarked that it had become clear that "as the scientific side of Industry developed, there should be greater and greater need for workers who were well trained in mathematics, physics and engineering."[26] Following discussion, it was decided that a circular be sent to parents and high school principals informing them that:

There is a demand at the present time for young men of good ability, well above the average, who have received both at school and university a sound and advanced training in mathematics and physics ... In the first place there are not sufficient Canadians to fill the professorships in mathematics and physics at Canadians universities ... This appeal is, therefore, made to parents, schoolmasters and to pupils and students directing attention to a profession with a promising future, which is almost unique in the fact that the demand is in excess of the supply, open only to men of marked ability who have been fortunate enough to come in contact with an inspiring teacher who is capable of directing them from an early age along the famous paths trod by the great thinkers of the ages.

This was especially true for physics, for in chemistry and biology "the organization is already complete and the number of students entering the professions is for the most part equal to the demand."[27] Until now, members of the committee had been able to speak out on physics education only on an individual basis. Henceforth, backed by the authority of the NRC, they were able to express their views in a collective manner and thus to contribute to the consolidation of their discipline.

Throughout its existence, the committee showed continued interest in physics teaching at both high school and university. Seven years after publication of the first circular, the committee considered publishing a bulletin under the auspices of the NRC "setting forth the need for young physicists in Canada and the opportunities awaiting them." However, it was agreed that "it would be difficult to do much work through the high schools." Nonetheless, T. Alty of the University of Saskatchewan remained convinced that "the greatest revision of courses [in physics] was necessary in the high schools rather than in the universities." Boyle, in contrast, pointed to the fact that "two eminent physicists in Canada [are] now deans of graduate schools at McGill and Toronto"[28] as an indication of the improved possibilities for the expansion of physics in the universities. In any event, with education falling within provincial jurisdiction, much of the discussion over standards and quality of science education on a Canada-wide basis was destined to go nowhere. The National Conference of Canadian Universities had met every two years since 1915 and had as yet been unable to standardize higher education across Canada.[29]

The Associate Committee's promotion of physics was conveniently integrated into the NRC's mission as a whole, which was development of industrial research in Canada. Speaking of its work in science education, the committee's annual report expressed the hope that "in this way, as well as in others, [we will] add to the supply of research workers that are likely to be required in the near future to develop through their efforts the industries of Canada, *and to meet the needs of the scientific life of the country generally.*"[30] The committee's members thus placed themselves in an ideal position. By emphasizing industry's need for physicists they were simultaneously able to promote growth of the physics discipline. As we saw in chapter 2, members of the NRC used a similar strategy to promote scientific research in Canadian universities.

Generally speaking, members of the committee made reference to the needs of the discipline rather than of individuals. The process of "need identification" could be carried out only by the discipline's "representatives"; ultimately, they would be called on to speak in the name of individuals who had not yet chosen to become physicists and who would be unlikely to do so until they perceived employment opportunities or ways of becoming "physicists." As it pursued its recruitment campaign in the schools, the committee also scrutinized the "output" of the training process in an attempt to assure that future physicists would have jobs. However, while the quality of science teaching was a constant subject of discussion, the role of the physicist in industry rarely caught the attention of the committee, which discussed the issue only twice in the course of its existence.

During the second meeting, committee members were supposed to consider the "practical means of introducing physicists into industries," but discussion was limited to the reading of a suggestion, made in absentia by Louis Bourgoin of the Ecole Polytechnique of Montreal, proposing a publicity campaign to rectify the fact that "the part that the physicist played [in industry had been] lost sight of and appeared as the engineer's work."[31] Thinking perhaps that the circular prepared for teachers and parents of high-school students contained a similar message, the committee made no decision on this issue.

The problem of the physicist's job market returned nine years later, during the committee's final meeting. A.L. Clark, of Queen's, complained that many Canadian physics graduates emigrated to the United States, never to return. According to Clark, the situation was, in large measure, caused by "the fact that many Canadian industries were branches of American concerns and were dependent for research on the parent companies in the United States."[32] One member suggested that an inquiry be conducted among Canadian industrialists. Clark replied that such an investigation had been carried out by the Canadian Manufacturers' Association, which had concluded that "little research work was being done by such companies in Canada."[33] It was therefore decided to recommend that the NRC direct the attention of the dominion government "to the desirability of having foreign companies with Canadian branches do as much as possible of their research work in this country."[34] As this was the committee's last intervention, its members' sudden understanding of the relation between economic dependence and the development of industrial research gave rise to no further action. The committee was not officially dissolved, but McLennan's departure in 1931 was a fatal blow.[35]

Compared to problems related to physics teaching, those concerning industrial research constituted a relatively minor preoccupation for the committee. Louis Bourgoin was the first to raise the issue of physicists in industry. As a professor at the Ecole Polytechnique, he was in a good position to understand this problem, which was even more acute for French Canadians. The committee was composed primarily of university professors engaged in pure science, and their major concern was the teaching of their discipline. Since no shortage of physicists in Canada was in view, there seemed to be no reason to worry about jobs. Between 1921 and 1928 Toronto and McGill granted eighteen PHDs in physics and Canadian universities hired fourteen new physics professors. In 1919, academic prospects were so good for physics graduates that A.S. Eve believed that there would be a shortage of researchers if the universities did not make an effort to train more graduates.[36]

CHAPTER SEVEN

Profession or Discipline?

During the 1930s there was a noticeable increase in the number of master's and doctoral graduates, as well as of those who held an honours bachelors degree in physics and who had found employment in Canadian industry. According to a study carried out by Statistics Canada in 1939, in collaboration with the NRC, eight physicists were employed in provincial research laboratories, thirty-one in dominion laboratories, seven in municipal laboratories, and thirty-eight in industrial laboratories.[1] During the Second World War, the number of positions grew considerably. At the NRC, for example, only fifteen physicists had positions as researchers in 1939, while, in 1946, the physics division had forty researchers, to which must be added another thirty at the atomic energy laboratory at Chalk River.[2] It is difficult to obtain precise figures on the number of physicists employed in industry at the end of the war. Clearly, however, the diversification of the positions they occupied created a problem in defining the social image of the physicist.

Whereas in the 1920s the problems of the physics discipline were mainly academic (publication, research grants, student recruitment), during the period 1930–45 a new group and a new set of problems emerged within the heretofore homogeneous group of physicists. Henceforth the industrially oriented "professional physicist" would seek recognition both within physics and within Canadian society as a whole. But, before detailing the activities of the founders of the Canadian Association of Professional Physicists, we must look at the historical context within which this organization emerged.

SCIENTISTS, ENGINEERS, AND COLLECTIVE BARGAINING

When physicists, or at least some of them, re-emerged into the public sphere toward the end of the Second World War, it was with a new discourse: having worked in industry, next to engineers, they had learned that outside the university the "physicist" received no recognition and that in industry engineers occupied the entire terrain and had appropriated the knowledge that they had believed that they alone were capable of exchanging for a salary.

This awakening of "industrial physicists," who claimed the same privileges as engineers (who were recognized as "professionals"), took place in a context similar to that which in Britain had given rise to the Institute of Physics and the National Union of Scientific Workers twenty-five years earlier.[3]

In Canada, these new associations were called the Canadian Council of Professional Engineers and Scientists (CCPES), the Canadian Association of Scientific Workers (CASCW), and, finally, the Canadian Association of Professional Physicists (CAPP) – later the Canadian Association of Physicists (CAP). Although I cannot here give a complete history of these associations and the relations among them, these organizations crystallized about proposed federal legislation concerning collective bargaining.[4]

Usually within provincial jurisdiction, labour relations at the beginning of the war became a problem of national interest, given the dangers of work-stoppages. Having begun in 1929, federal intervention in this area culminated in February 1944 with an order-in-council that established compulsory collective bargaining and arbitration administered by a Bureau of Labour Relations.[5] The proposed legislation, which was to become law after the war, made Canadian engineers and scientists ask: should we, like physicians and lawyers, ask to be excluded from the law, or should we participate in collective bargaining? Thus posed, the question could not help but incite the creation of different associations, the "delegates" and "representatives" of which would speak in the name and interest of the respective groups.

In January 1945, a dozen engineering and scientific associations formed a coalition and founded the Canadian Council of Professional Engineers and Scientists (CCPES), with the aim of presenting a united front to the federal minister of labour and to take a position on collective bargaining in the name of 40,000 engineers, chemists, surveyors, and architects. However, a number of scientists, who identified themselves as employees and whose activities, unlike those of the engineers, had no legal recognition, saw the proposed legislation (which was to replace the order-in-council) as an opportunity to gain recognition finally as "scientific workers," with the right to negotiate collectively their working conditions.

Politically to the left of the engineering associations (and unlike the CCPES, which existed only because of the proposed legislation), the Canadian Association of Scientific Workers (CASCW) had given itself the same objectives as the British Association of Scientific Workers presented in 1939 by John D. Bernal in *The Social Function of Science*. "The basis of the Association is now recognized to be twofold: one, professional and individual concern with preserving and improving the conditions of employment of its members, and establishing the status of the "scientific worker" as in some way similar to that of the doctor or the lawyer; the other concern is with the whole position of science in society. The two are closely linked."[6] If the first objective could easily gain the consent of a large number of Canadian scientists, the second, which had a "socialist" hue, would have stopped many scientists from becoming members. Despite a promising beginning, the CASCW was put on the spot by the report of the Royal Commission of Enquiry into the Gouzenko Affair, published in July 1946: it branded the association as pro-communist.[7]

While the CASCW supported the proposed labour legislation, the CCPES, with which the former did not wish to be associated, found it difficult to manage the different opinions expressed within its association, resulting from the fact that not all members perceived themselves as "employees." Already, in June 1945, the Engineering Institute of Canada (EIC), which opposed the legislation, denounced the CCPES, which was seeking a solution that would allow engineers and scientists to be "brought within the scope of the proposed Labour Code."[8] Divided, the CCPES was unable to take a position, and the EIC, which presented itself as the "dominant organization of Engineers in Canada," won.

When the law was promulgated, on 30 June 1948, engineers found themselves among the lawyers, physicians, and architects as professionals unable to undertake collective bargaining. In the course of the operation, the notion of "scientific worker" had its rise and decline, which dismayed the directors of the CASCW and some directors of the CCPES. Commenting on adoption of the legislation, the Montreal branch of the CASCW expressed its surprise and disagreement: "Among the changes was a totally unexpected one which nullified the recognition of the right of scientific workers to bargain collectively ... a stand for which the CASCW fought so hard in 1944. In effect, at present, scientific workers are not granted the protection and rights of the federal labour legislation. Scientific workers in the Province of Quebec, in particular, have no possible recourse to law in matters of collective bargaining."[9]

Having no further reason for existence, the CCPES disbanded in 1949. The CASCW sunk out of sight, its members having probably followed the editor of the Montreal branch's *Bulletin*, the mathematical physicist R.W. Wallace, who resurfaced at the end of 1949 as a member of the editorial committee of the *Bulletin* of the Canadian Association of Physicists.[10]

It was in this context of organizational fever – incited by an order-in-council on collective bargaining rights for employees of Canadian industries – that the idea of a Canadian association of professional physicists emerged. Set up within a crown corporation on the verge of shutting down, Research Enterprises Ltd (REL), the projected association sought to assure physicists in industry a position comparable to that of engineers. As Fred E. Coombs, president of the organizing committee of the CAPP, wrote to the president of REL: "Late in 1944, a few of your employees at REL were discussing some of their problems, and on this occasion the discussions concerned the manner in which physicists were overshadowed by engineers in Canadian industry."[11]

After having consulted scientists such as Robert W. Boyle of the NRC; E.F. Burton, head of the physics department of the University of Toronto and a director of REL; C.J. Mackenzie, president of the NRC; and the directors of the Chemical Institute of Canada and the Canadian Manufacturers' Association, the group, composed for the most part of industrial physicists, felt ready to launch the association officially by writing letters to approximately 500 Canadian physicists (or at least those recognized as such by the organizing committee).[12]

Dated 1 April 1945, the introductory letter clearly indicated that the organizers intended to concentrate on problems related to physicists in Canadian industry: "This letter has been prepared to bring to your attention the ideas of a group that feels that a Canadian Association of Professional Physicists should be formed among those whose employment depends on the utilization of the science of physics."[13] As it was impossible to exclude university physicists, the letter continued: "Since certain very creditable contributions have been made to technical advances in munitions by the so-called 'academic' physicists, the matter being discussed herein has application to all physicists rather than just to those directly employed in industry."

The authors recalled that if physics had been traditionally considered a pure science that did not deal with practical applications, the situation had now changed, since "in recent years the rapid pace of scientific developments, especially in electronics, has caused physicists to enter industry in increasing numbers." Finally, the recent establishment of the CCPES – which included several professional organizations and which, according to the authors, "will no doubt be responsible for some co-ordination and leadership of the utmost significance in the post-war era" – added to the urgency of the situation, for, without an organization like the CAPP, physicists' interests would not be represented within the CCPES.

Since the organizing committee wanted above all to secure physicists a status similar to engineers, it drew up a provisional charter, stating that the objectives of the CAPP were: "a) to secure recognition of Professional status for the physicists; b) to protect the status of the profession by setting professional standards, and by controlling admission to the profession."[14] Discussion polarized over these two articles, which revealed two opposing tendencies within the group of physicists.

The first responses received by the organizing committee showed immediately that industrial and academic physicists did not have the same conception of the organization. Employed by Imperial Oil in Sarnia, Peter J. Sandiford, a specialist in Raman spectroscopy, was well placed to understand the committee's point of view. He wrote enthusiastically to the president, Fred E. Coombs:

For some months I have been casting around in my mind with a view to finding a way to organize Canadian physicists into a national society ... For over a year now I have been a member of the executive of the local branch of the Canadian Chemical Association ... During this period I have been impressed with the benefit that such an organization brings both to its members and to society in general. I have also been acutely aware of the disorganized and aimless situation in which the Canadian physicists find themselves at present.[15]

At the other extreme was R.C. Dearle of the University of Western Ontario, who felt that by taking into account only industrial physicists the CAPP depreciated university physicists. Like many others, Dearle pointed out: "The members of the organization committee are not sufficiently well known to physicists at large in the Dominion to establish confidence in this new venture. There should be included the names of some who have a national reputation."[16]

By focusing on problems relating to physicists in industry, the objectives of the association failed to attract the attention of the most prominent scientists, who could have come only from the university. It was only after the first national meeting, at which such well-known scientists as J.D. Cockcroft, J.S. Foster, and R.W. Boyle, gave talks, that the problem of credibility would be overcome and university physicists became the majority of members. As we shall see, this led to major modifications in the association's objectives.

Following Dearle's comments, his colleague R.L. Allen suggested that the CAPP's name be changed to the Canadian Institute of Physics and proposed a new constitution whose objectives were "a) the advance of physics in all its branches; b) the furthering of the existence of physics in general; c) the establishment of professional standards."[17] Naturally, acceptance of these propositions would lead to another kind of organization than that envisioned by the promoters of the CAPP, whose central objective had been to defend the interests of physicists employed in industry and not to create a simple meeting place where researchers could discuss research projects.

In his reply to Allen, Coombs took care to distinguish between these two types of organization:

We have considered that an Institute is primary concerned with a science, whereas an Association is primary interested in scientists. The Engineering Institute of Canada is concerned chiefly with the science, and the Association of Professional Engineers is concerned chiefly with the scientists. In pure science the Royal Society of Canada acts as the Institute, and is divided into several sections for the various pure sciences ... The physicists have the interests of their science represented by the mathematics, physics, and astronomy sections of the Royal Society of Canada, and the interests of their scientists are unrepresented. This is the reason that we feel that a professional association of physicists is required ... Most of the younger physicists who are meeting present-day problems consider that such a group is absolutely necessary in order to protect their professional interests.[18]

For the founders of the CAPP, it was clearly a question of controlling, in a manner similar to engineers, access to the practice of physics. As Coombs specified in a letter to Sandiford: "Our plan is to establish a National Association ... to perform the functions corresponding to the Engineering Institute of Canada. Then, at a later date, we would attempt to form provincial chapters of this parent Association into semi-autonomous bodies capable of controlling the profession within the province."[19]

The idea of controlling access to the "profession" was anathema not only to academics. Midway between academics and industrial physicists were employees of research institutions, such as the NRC, who hoped to increase their salary but had no interest in controlling the category "physicist." Noticing the reticence of NRC physicists to join the CAPP, the president asked one of them (termed "representative") to specify his grievances. While claiming not to represent his colleagues, Alex Ferguson answered that he did not wish to see the term "physicist" controlled by an organization like the term "engineer" was. He none the less believed that an association was needed to defend scientists' interests, particularly with regard to salaries, and that the CASCW, "being larger than physics, will have more weight." Finally, Ferguson thought that "any people who are likely to employ physicists will know fairly well what they want."[20] For the secretary of the CAPP, this latter opinion demonstrated the exceptional nature of NRC employees' position. In his reply, he added: "Those of us who have worked in industry can assure you that such is not the case there. The industrialist, if he has heard the term physicists at all, certainly has no idea what it means, or what such a man is capable of doing other than making atomic bombs, except perhaps in very rare cases."[21]

In order to make industrialists more aware of the importance of hiring physicists, the organizing committee asked J.J. Brown, a former professor of English who had become head of the Division of Technical Publications at REL, to write an article on the relations between physics and industry. Published in the August 1945 issue of *Industrial Canada*, the article sought to attract attention to the creation of the CAPP and to show that "the physicist is not only a person capable of making abstruse calculations in a laboratory, but a practical research worker in industry, following a project from its first conception in the laboratory, through the manufacturing process in the factory, to the actual operation in the field."[22]

It was a question of "deconstructing" the Einsteinian image of the physicist, distracted and incapable of adapting to the concrete realities of Canadian industry, and substituting for it the image of a "professional physicist," capable of competing with engineers "in the field." By recalling the multiple contributions physicists had made during the war (such as radar and acoustic detection of submarines), the new techniques they had perfected for industry (such as x-rays applied to metallurgy), and the new industrial perspectives opened up by recent work (as in polaroid and fluorescent lighting), the author sought to convince readers that if circumstances had indeed forced industry to use physicists and had obliged "academic physicists" "to pay more attention to the needs of industry," this co-operation "should be self-consciously continued in peacetime. Physicists and other scientists should be encouraged to remain in Canada by the inducement of salaries commensurate with the years of training necessary to become a physicist, and with the difficulty of the work."[23]

The attention paid to relations between physicists and industrialists evidently reflected the interests of the association's directors. In fact, of the eleven members of the first council, there was only one academic physicist, A.D. Misener of the University of Toronto. The president, Fred E. Coombs, and the secretary, N.J. Abbott, as well as four of the six members of the executive committee, were employees of REL. The sixty-seven responses received by July indicated that those in the university milieu were less eager to join the new association than workers in industry and government. There were only twenty university members, of whom half came from McGill and Toronto, and five NRC employees, as opposed to thirteen from the Meteorological Office of Canada, twelve from REL, and fifteen from assorted companies (RCA, Kodak, Ford,

Imperial Oil, and so on). One year later, academics still accounted for only 40 per cent of the 134 members. They formed a majority only by December 1947, when they accounted for 65 per cent of members. As we shall see, the reversal of proportions had an important effect on the future of the association.

As the CAPP's initial project had been to obtain recognition of specific rights for physicists, the association was supposed to have led to the granting of a professional charter. There were two obstacles to realization of this objective. The first was constitutional, and the second "demographic," so to speak.

From a legal point of view, management of the professions and labour relations fell within provincial, not federal jurisdiction. Already in February 1945, Boyle had written to the organizers: "If an Association of professional physicists were formed ... its charter would fall under the jurisdiction of the Provinces rather than the Federal Government, in accordance with the stipulations of the British North America Act."[24] Boyle was well abreast of the issue, as he had been an active member and president of the Association of Professional Engineers of Alberta.[25] Now, in 1949, although the organizers were aware of the two levels of jurisdiction, they none the less submitted their charter to the federal government, even though the association's objectives made explicit reference to control of the practice of the physics profession. Naturally, the demand was refused.

The association's president, D.C. Rose, then met with legal counsel, who advised the association that unless it submitted a private member's bill, which would be costly and risky, to obtain federal incorporation, it would have to remove "all mention of professional standards from [the] constitution and re-apply for a Dominion charter as a scientific organization." After four years' work, the executive thus concluded: "A charter as a scientific organization would be better than no charter at all. It would provide legal protection for the executive and members of the Association, and would give to the Association a recognized status which, it was hoped, would further its growth and increase its ability to assist its members."[26]

The council of the association therefore rewrote the constitution to conform to federal law. It removed the articles claiming "recognition" and "protection" of the "professional" status of the physicist and substituted more disinterested objectives. Henceforth, the association was to devote itself to "a) further the advancement of the science of physics; b) to promote the use of physical discoveries in the interest of mankind."[27]

The new charter was officially accepted by the federal government on 9 June 1951, but the forced modifications had revived tensions between the industrial physicists, partisans of a professional association, and the academics, who had a more "disinterested" view of their activity, as they had not been forced to confront engineers in Canadian industry.[28] Anticipating the reaction of industrial physicists, Elizabeth J. Allin, professor at the University of Toronto and secretary of the association, noted: "Disappointment at the necessity of removing all explicit reference to professional activity from the constitution should be tempered by the expectation that the activities of the Association will hasten the formation of provincial organizations with powers in professional matters, since they will bring physicists together and provide opportunities for them to discuss their mutual problems."[29]

CAP: WHAT'S IN A NAME?

The federal charter of 1951 completely changed the nature of the association, which now conformed to the wishes of those who in 1945 had suggested a constitution of the kind that now became official. The "academic" fraction of the "community" of physicists had thus won all down the line, for it had also obtained a change in name. In early 1947, following an inquiry among members that indicated that the majority opposed use of the term "professional" in the association's name, the council had decided to change the association's name to the more neutral Canadian Association of Physicists (CAP).[30] In 1947, this defeat may have appeared minor to the industrial faction at the origin of CAP, for it did not change the constitution.[31] Two years later, however, the new charter called into question the primary objective of the association's founders and provoked important reactions.

In a letter published in December 1950 in the association's bulletin, C. Barnes opened the debate on the usefulness of an organization that had been unable to obtain a charter incorporating the objectives for which it had been created – namely, protection of the interests of industrial physicists – and suggested that CAP simply be dissolved.[32] In the bulletin's subsequent issue, one of the organization's pioneers, Peter J. Sandiford, then working for a Toronto hydroelectric company, denounced the academics' domination of an organization which had been set up to unite physicists working in the industrial milieu and who had common interests to defend:

Much to my personal regret, as one who helped establish C.A.P., I find too much truth in Dr. Barnes letter in the previous bulletin. Professional societies in engineering, medicine and the law are possible because the majority of the profession are engaged in practicing it. In physics on the other hand we find very few physicists. There are teachers whose common interest lies with their university colleagues in fields ranging from Arabic to Zoology. There are civil servants whose interests lie with the postmen and customs officials. But there are few physicists earning their living by selling their knowledge and judgement in the field of physics ...[33]

As the annual meeting did not justify the existence of CAP, Sandiford concluded: "Either C.A.P. should drop the academic people, re-name itself the Association of Industrial Physicists of Ontario (Quebec, etc., etc.) and attack afresh the problems it was created to solve, or else should reconsider the offer once made to it by the British Institute of Physics, that we become the Canadian branch of that organization."

In effect, industrial physicists had no choice but to begin again at zero, since, as these lines were being written, the association was clearly dominated by academics. In 1950, industrial physicists no longer monopolized the positions of president, vice-president, secretary, and treasurer, as they had in the preceding years.

In its reply to Barnes's letter, the executive of CAP presented its view of the association's objectives. After having recalled that a professional association was a provincial matter, the executive noted:

This country needs an Institute of Physics. The Royal Society of Canada's Section III cannot be expected to perform all the duties of such an Institute at the same time that it proceeds with its primary purpose of epitomizing thought in all the physical sciences in Canada. In its role of physical society, CAP is in a strong position to foster scholarship. It does so, now by its annual congresses ... by its annual review of physics in Canada. The next constructive step might very well be the inauguration of inter-university lectures, enabling each research centre to learn more of activities at the other places, more of the men at the other places.[34]

These fundamental objectives were far removed from what F.E. Coombs had announced five years earlier. Faced with this purely academic conception of the association, physicists employed in industry and in governmental institutions (both were concentrated in Ontario) made a second stab at creating a professional association: the Ontario Association of Physicists. Among the twenty-two members of the organizing committee were five of the thirteen founders of CAP and ten academic physicists. Their objective remained the same: "to establish professional status, through the creation of an Association and by reserving the name "professional physicists" for members of the Association. Only by law can professional status be achieved."[35]

However, following a survey of the situation and consultation with the executive of the Association of Professional Engineers of Ontario, the committee concluded that the costs of preparing a professional charter ($10,000 to $15,000) and the expenses entailed in maintaining such an association would result in membership dues that would be far too high, given the relatively few potential members. Confronted with this situation, "the committee unanimously felt that this very effectively ended their consideration of a separate licensing body."[36] Finally, a modification of the engineers' charter to include physicists would "put the whole act in jeopardy," with the result that the association was not created and physicists did not obtain their "professional" rights, unless as individuals they managed to join an engineering association.

This second failure marked the end of the project of professionalization of physicists. This project was a strategy by industrial physicists – a group dominated by engineers – to control an occupation in order to assure themselves a place in a market which in industrial milieu was monopolized by engineers. Few in number, industrial physicists had to ally themselves with university physicists. The latter rapidly marginalized the former within their own organization. In addition to the "tactical error," of having allied themselves with academics, the founders of the CAPP made a fatal "strategic error": they had invested four years in preparation of a federal charter, while the whole project should have begun at the provincial level, in particular, in Ontario.

Unresolved, the professional status of the Canadian physicist resurfaced on several future occasions. In 1983, CAP formed a committee on professions which suggested creation of professional societies including all Canadian scientists. In its report tabled in October 1987, the committee insisted on the importance of such action; otherwise "*the practice of the professional scientist will continue to be eroded ...* by licensure regimes, such as that of the Professional Engineer, and by certification regimes, such as technologists."[37] As the association was still dominated by university physicists, the committee drew attention to the fact that "another consequence, that will be seen by university staff, is that *students will seek "professional" areas rather than science.*" In seeking to ally themselves with other scientific disciplines in order to give greater weight to their demands, the physicists had made the solution to their problem even more difficult. In any event, only the future will tell if the physicists will ever succeed in constructing a profession either alone or in concert with other scientists.

CHAPTER EIGHT

Representing the Discipline

Unable to charter a profession, the directors of the Canadian Association of Physicists redefined their objectives after 1951 and conceived of CAP as a site for information exchange among researchers. This turnabout had been simple: because the majority of the membership was academic, members were sensitive more to issues in communication in research than to problems in labour relations, negotiations which concerned only a minority. Moreover, the most concrete realization of the association since 1945 had been the annual meeting, bringing together Canadian physicists to discuss the latest developments in their discipline.

A NEW MEETING PLACE

In September 1945, the organizing committee of the then Canadian Association of Professional Physicists (CAPP) was already preparing a conference where discoveries made during the war could be discussed openly for the first time.[1] According to organizers, a meeting between physicists would reopen networks of communication and discussion interrupted by war. However, as these kinds of objectives are generally pursued by a scientific association, the fact that the meeting was organized by CAPP resulted in some confusion about the group's objectives. In reply to a request for financial aid, the president of the NRC replied: "If your physicists are organizing on a professional basis like the other professional bodies, the Research Council probably could not give any financial support to meeting for, as I have said, the policy of the Council has been to restrict grants to the scientific congresses and similar general meetings of various bodies."[2]

In practice, the meeting resembled the usual scientific congress, and only the claim for professional status for physicists confused the situation. Nonetheless, the meeting took place as planned at the McLennan Laboratory of the University of Toronto in May 1946 and brought together 200 physicists to hear invited speakers on nuclear physics, meteorology, and the physics of radar. These three areas had a common characteristic: they had evolved the most during the war. "Noblesse oblige," the opening talk, devoted to nuclear physics, was given by John D. Cockcroft, director 1944–46 of the nuclear research laboratory in Montreal and at Chalk River. Cockcroft presented the basic theory of nuclear fission and its commercial applications in a nuclear generator. He had directed construction of a natural uranium reactor and of a heavy water reactor, which on 5 December 1945 became the first to function outside the United States.[3]

John S. Foster of McGill had worked from 1941 to 1944 at MIT's Radiation Laboratory on radar equipment. As the meeting took place in the shadow of the Gouzenko Affair, Foster, before accepting the invitation, had assured himself that the CAPP was not a communist organization.[4] As the development of radar had also stimulated research in electromagnetism, William H. Watson, Foster's former colleague at McGill, then director of the mathematics department at Saskatchewan, presented the electromagnetic theory of wave guides and resonant cavities. Finally, P.D. McTaggart-Cowan of the Meteorological Office of Canada presented the new techniques of temperature prediction developed during the war.[5]

Given the success of this first meeting, a second was organized the following year at the University of Western Ontario, which included a session where physicists could present work-in-progress, thus adopting the model of the American Physical Society, where Canadian physicists often presented their work. Following this second meeting, which attracted some 120 participants, the directors of the association (just renamed CAP) concluded: "At the present stage of its growth, one of the best contributions the c.a.p. can make is the organization of an annual congress, bringing together Canadian physicists of widely dispersed interests. The provision of an interesting program, and opportunities to meet over lunch table stimulates those who attend, and is an important factor in keeping the Association alive."[6]

Even though CAP's objectives remained professional until 1949, in practice it functioned as a traditional scientific society. As such, it initially overlapped with the Royal Society of Canada, which also held an annual meeting where physicists could present their research results. However, CAP's more "democratic" functioning made the participation of young researchers easier, and the two meetings quickly became considered complementary.[7]

The federal government's refusal in 1949 to grant professional standing to the association was the ideal occasion for the new organizers to dissipate the ambiguity that hung over CAP. During his presidential address at the May 1950 meeting at McMaster, D.C. Rose exposed his conception of the association. While recognizing that it was legitimate for physicists employed in industry to seek to protect themselves from exploitation, he none the less believed that "an organization of men who are responsible for the advancement of physics must have as one of its primary objectives the exchange of information ... Those who hold strong views on making the c.a.p. an association mainly to set professional standards must bear this in mind. The major portion of the annual congress or meetings of any such organization must clearly be for the exchange of scientific information and discussion of recent developments."[8]

For Rose, who was indeed presenting the academic viewpoint, the "professional" aspect of a physicist's work referred less to occupational control than to the physicist's "moral" responsibility toward applications of his research : "A consciousness of responsibilities of this nature is to my mind one of the reasons that will make an organization of physicists a successful *professional* organization."

Pointing out the medical profession as the finest example of a professional association, he added: "I have not studied the details of their code of ethics ... but I am sure their organization is based on the ideals of service to humanity in the use of their science." Rose thereby inverted the concept of the profession, which, until then, had served as the basis of the association: whereas the first directors had seen in the medical profession a perfect example of the control of access to an occupation, Rose emphasized disinterested concern for humanity. This moral conception of the "profession" was, moreover, the only one convenient for a group of academics who had always opposed the more activist concept of the industrial physicists. As we saw above, industrial physicists did not easily accept the directors' new discourse and vainly attempted a second time to obtain a charter in Ontario.

In the early 1950s, therefore, physicists abandoned the theme of "professionalization," and the association concentrated henceforth on problems of the discipline, thus returning to questions not touched since dissolution in 1931 of the nrc's Associate Committee.

THE USES OF REPRESENTATION

Although the annual meeting, which visited in turn the main Canadian university campuses, constituted cap's main – because most visible – form of existence for many physicists, "representatives" gave life, between meetings, to cap by making decisions in the name of Canadian physicists.

As the name indicates, the executive committee, composed of a president, vice-president, secretary, treasurer, and three directors, all named for a period of one year, was occupied with the "daily" functioning of CAP, assuring application of the decisions taken by the Council (composed of "regional" representatives and which defined general policy) or by the members meeting annually in the general assembly. As we saw at the end of chapter 5, one of the first exercises in representation was to contact the editor-in-chief of the *Canadian Journal of Research* in June 1951 to demand representation for CAP on the editorial committee. As CAP was unable to produce its own journal, unlike the American Institute of Physics, this understanding was the simplest way of acquiring a scholarly journal. It was hoped that, as the years went by, "C.A.P. will obtain influence in the editorial board of this publication."[9] From January 1952 on, therefore, members received a journal with a new name, the *Canadian Journal of Physics*, which could only have strengthened their identity as "Canadian physicists" and encouraged them to publish in their "national" journal.[10]

In 1949, in order to develop Canadian physicists' "esprit de corps," CAP began publishing an annual review, *Physics in Canada*, which presented research in physics carried out in Canadian universities, industry, and government institutions. The *Bulletin* continued to be published four times a year. It contained information about the association, the texts of some presentations made at the annual meeting, and information about Canadian physicists' current activities. Beginning in 1952 the two journals were combined into a single *Physics in Canada*, which appeared quarterly. Each issue contained, in addition to several articles, a section on the movements of Canadian physicists. There was thus created, in spite of distance, a feeling of belonging to the "family" of Canadian physicists. As an example, a physicist in New Brunswick who had read the winter 1952 issue could say to his colleague: "Did you know that Dearle has just been named director of the Ontario Research Foundation?" Even if he or she had never met these people personally, names like Shrum, Volkoff, Kerwin, and Katz thus became more familiar.

Although francophone physicists had made their presence felt within the association by the late 1940s – the 1949 meeting was held at Laval – the review did not become bilingual until the 1960s.[11] In the spring of 1962, the title *Physics in Canada* was still unilingual, but a new French subtitle had appeared: "Bulletin de l'Association canadienne des physiciens." In 1965, the French title *La physique au Canada* was added to the cover. It seems it was necessary to wait for the "Quiet Revolution" for anglophone physicists to realize that Canada also had a francophone community.

The possibility of awarding an annual medal in recognition of important work done by a Canadian physicist naturally appeared quite early on the agenda of the executive committee. In May 1953, it formed a committee to study the question. Although the committee suggested

establishment of two medals – one for research and one for teaching –
the executive thought it wiser to restrict itself to just one medal, "For
Achievement in Physics," to be awarded for either activity.[12]

Since 1925, the Royal Society of Canada had awarded the Flavelle
Medal in recognition of important scientific achievements by Canadian
scientists. Until 1943, the awarding of the medal had been the work of
a committee composed of representatives of all the scientific sections.
In that period, three Canadian physicists had received the Flavelle Medal:
John C. McLennan in 1926, Louis V. King in 1934, and Robert W. Boyle
in 1940. In 1943, the institution of the Henry Marshall Tory Medal
allowed for greater "specialization" of awards. The new medal honoured
work in disciplines represented in section III (astronomy, chemistry,
mathematics, physics), while the Flavelle Medal was given to disciplines
covered by section V (biology).[13] Thus, Frank Allen received the Tory
Medal in 1944, followed by John S. Foster in 1946 and Elie F. Burton
the following year. Subsequently, the Tory Medal was awarded every
two years.

When a new medal is created, there is generally a "waiting list" which
is difficult not to take into account. The first CAP medals were therefore
awarded to veterans of physics in Canada who had trained many of the
physicists active in the association during the 1950s. Thus, J.A. Gray,
the first medal-winner, in 1956, had done his most important work on
x-rays during the 1920s, for which he had been elected to the Royal
Society of London in 1932. Gray had retired in 1952 at the age of sixty-
eight and the medal was awarded him as much in recognition of the
importance of his work as for his contribution to the training of more
than fifty physicists, many of whom held "important positions in
universities and atomic energy projects in Canada and the United
States."[14] The following year the medal was awarded to an active
researcher, Gerhard Herzberg, director of the physics division of the
NRC and internationally recognized for his work in atomic and molecular
spectroscopy, for which he obtained the Nobel Prize in chemistry in
1971.[15] In 1958, it was John S. Foster's turn, for his work on the Stark
effect carried out in the 1920s, which had also earned him the Levy
Medal of the Franklin Institute in the United States. Like Gray, Foster
was honoured for his services "to physics and to physicists in Canada
during the past forty years."[16]

Even though the CAP medal was not as prestigious as the Tory Medal,
it succeeded in accumulating prestige as it was first awarded to physicists
who were already recognized and who brought more prestige to the
medal than the medal gave to them. Once this "primitive accumulation"
of prestige had been achieved, the gold medal "For Achievement
in Physics" could fulfil its role and "reward" the efforts of younger

researchers who had received their doctorate in the 1930s (such as R.W. Sargent, in 1959, and H.L. Welsh, in 1961) or at the beginning of the 1940s (such as D.W.C. Macdonald, fifth medallist, in 1960).[17]

RECRUITING NEW PHYSICISTS

If publishing of a quarterly bulletin, holding an annual meeting, and awarding a medal permitted creation of a "community of Canadian physicists" and the management of its internal affairs, another type of action was necessary to regulate relations with the outside world – Canadian society and its constituent groups. Representing themselves to other groups not only served physicists' interests but contributed to the production of their identity, for the success of their representations also depended on the image they presented and on what was reflected back to them. Thus, with the recurrent theme of the quality of science education in high schools, the association presented a sharply different image from the one that the first directors of the defunct CAPP attempted to circulate.

Like the members of the NRC's Associate Committee on Physics and Engineering Physics twenty-five years earlier, the directors of CAP clearly perceived the importance of high-school science education for the long-term reproduction of their discipline. According to Gordon Shrum, president in 1953: "A teacher who has a very limited knowledge of the subject can never present physics to his pupils as the greatest intellectual adventure of our age … Consequently he will not win for physics those creative, and imaginative members of his class who are at home in the world of quantity, number and measurement."[18]

CAP was thus forced to act at the level of both students and teachers in order to orient "promising high schools students to careers in physics."[19] Starting in the summer of 1955, meetings with high-school science teachers and students were organized by a number of university physics departments across Canada in order to discuss career possibilities in physics.[20] A brochure titled *Physics in Canada: A Career and a Vocation* was sent out the next fall to attract the attention of high-school graduates in search of a career.[21] Even though the brochure mentioned that the majority of physics graduates found employment either in industry or within a governmental organization, physics was none the less presented as "one of the most satisfying careers that a young Canadian can select." The physicist himself was endowed with a somewhat mystical image: "In this quest for an understanding of nature the physicist is fulfilling the purpose of his creator: that Man should know Him, and understand His manifestations. Physics is a vocation."[22]

The brochure went so far as to specify the behaviour (expected or habitual, it wasn't clear which) of a physicist:

Does he [want] ... to change the face of nature, to harness and control natural forces? No physicists here ... for although they study these things they rarely actually do them. Does [he] feel an interest in his fellow man, and wish to make personal, intimate contributions to his welfare? A physician, a social scientist, possibly, but not a physicist ... Does he like to stay in the laboratory for weeks or years, realizing that his piece of experimental apparatus contains as many mysteries as the whole universe? Does he lose interest in an activity that becomes routine, and always want to be doing something new pushing into unexplored regions of knowledge? ... then he may well have been born a physicist.

This mythical (and mystical) conception of the physicist continually discovering new laws of nature, which ten years earlier J.J. Brown had attempted to dispel in the name of CAPP, thus returned as the centre of CAP's "publicity campaign." Already marginalized, industrial physicists must have felt quite foreign in the imaginary world of the academic physicist.

In order to carry out recruitment at the post-graduate level, in the spring of 1957 CAP inaugurated an annual competition for Canadian university graduates. Those with the best mark on the physics exam won a prize of $100.[23] In the fall of the same year, a system of itinerant speakers was set up to allow students in Canadian universities to become better acquainted with Canadian research in physics. During the 1957–58 school year, four physicists, at a cost of about $1,000, visited eighteen universities to present both their research and the association's activities.[24] Several years later, a similar program targeting high schools was instituted.[25]

Finally, through its committee on high-school education, CAP took a close interest in the American efforts to renew science education at this level, which gave rise to the Physical Science Study Group (PSSC) in 1956. Contact between the CAP and the directors of the PSSC enabled professors interested in science education to participate in a workshop held at MIT. This collaboration facilitated introduction of the new American science program into Canadian schools in the 1960s.[26]

The existence of a scientific association whose representatives were able to speak not only in the name of their members (which might appear corporatist) but also and above all in the name of the "discipline" (which appears more "disinterested") was necessary to co-ordinate the activities just described and to allow physicists to exist as a group. It became vital when the discipline entered the era of what is known as "Big Science," when the role of representatives became determinant in the decision-making process which defined "science policy."[27]

For Canadian physicists, entry into this new era took place just after the Second World War, when a large number of them returned to their universities after having worked on construction of a nuclear reactor, the fruit of collaboration between Canada and Britain.[28] With the war over, the federal government had created the Atomic Energy Control Board (AECB) to direct activities at the Chalk River scientific complex. Its mandate included, in addition to control of production and distribution of uranium and direction of research relative to atomic energy, a joint NRC/AECB grant program to develop nuclear physics in Canadian universities.

Requests for grants were not long in coming. At the second meeting of the AECB, members of the committee presided over by A.G.L. McNaughton received a request for financial aid from the University of Saskatchewan for construction of a 20-MeV electron accelerator. Several months later, McGill and British Columbia made similar requests, the first for a 100-MeV cyclotron and the second for a 3-MeV van de Graaff electrostatic accelerator. In granting a total of $150,000 for all these projects, the AECB and the NRC had definitively put Canadian physicists on the road to "Big Science."[29]

This equipment accelerated not only particles but also the rhythm of scientific discovery. The instruments aged quickly, and the construction time obliged physicists to plan long in advance the construction of new equipment (with the help of which they hoped to remain competitive in physics). Already in 1957 R.W. Sargent, of Queen's University, had written: "Those of us who were associated with the low energy accelerators were by 1955 beginning to realize that our machines were becoming obsolete and that with expanding knowledge it would become increasingly difficult to devise interesting experiments and to attract graduate students. We were and are seriously concerned with the questions: "What will our relative position be in 1960? What will it be in 1965?'"[30]

However, unlike low-energy accelerators, which could be managed by a single university, future machines implied costs and a kind of administration that made them impossible projects for one university alone. Only a collective effort would permit the construction of a high-energy (around 10-GeV) accelerator.

A project for a "national high-energy physics laboratory" thus took form. Naturally, none of the physicists at the origin of the idea was able to speak about it in his/her own name, and as the project was necessarily addressed to the federal government, the initiators could suggest only that their national association, CAP, be the sponsor. By taking in hand a project of interest to only a minority of its 485 members (in 1955), the association used the opportunity to act at a national level. Until then, the only "national" dimension of the association was the geographic distribution of its members. To present the country's government with a "national" project was an excellent way for CAP to raise its profile. In effect, in accepting even to consider the project, the government would recognize CAP's legitimate right to speak in the name of Canadian physicists, a form of recognition that its geographic representation did not automatically confer. Nuclear physicists themselves were happy to see the project presented in the name of all "Canadian physicists" and not just of "nuclear physicists."

At its 22 June 1955 meeting, the executive committee mandated Leon Katz, a nuclear physicist at the University of Saskatchewan and secretary of CAP, to sound out the opinions of the directors of the physics departments in Canadian universities.[31] Even though many of them wondered if Canada possessed the manpower necessary for such a project and feared that the universities would lose their best researchers, the majority of the fourteen department heads who replied favoured a high-energy accelerator.[32] On the basis of these opinions and with the support of the presidents of the NRC and the AECB, CAP's executive committee appointed Leon Katz, George Volkoff of the University of British Columbia, and Paul Lorrain of the Université de Montréal to do a feasibility study of the project, with the help of a $5,000 grant from the NRC.

The three physicists spent the summer of 1956 visiting the large accelerators in operation or in construction in Europe and the United States. Lorrain inspected synchrotrons at Saclay (2.5 GeV protons) and at Rome (1 GeV electrons) which were already in operation. After having met with fifty specialists in this growing area, the three wrote a report which they presented to CAP's executive committee in February 1957. They suggested construction of a traditional synchrotron-type proton accelerator with an energy of about 10 GeV. The energy level was determined "by recognizing the importance for a machine which will not be in operation before 1963 to produce both anti-nucleons and different kinds of heavy mesons in reasonable quantities."[33]

Now, the energy threshold for production of anti-particles from a proton pile striking a fixed target is 5.6 GeV. In order to produce a sufficient number of anti-particles, the energy level must be above this mark, whence the 10-GeV objective. Construction cost was estimated at around $20 million, spread out over five years, followed by annual operating costs of approximately $250,000. With regard to management of the project, the report remained vague and recommended only "that the policy-making body for this laboratory consist principally of scientists drawn from the Canadian universities."[34] As for the site of the laboratory, the highly political nature of the choice escaped nobody, and the committee limited itself to mentioning four possibilities in order of preference: Kingston, Vancouver, Montreal, and Edmonton.

After having analysed and accepted the report, the executive committee prepared a resolution, which B.W. Sargent and L.G. Elliot presented to CAP's general assembly on 14 June 1957 at the University of Ottawa. Adopted unanimously, the resolution recommended: "That the Executive of C.A.P. be authorized to present a brief through appropriate channels to the Government of Canada, recommending that a high energy project be set up in the form of a new laboratory centered around a high energy accelerator and located on or near the campus of one of the Canadian universities."[35]

As the John Diefenbaker's minority government was preparing for an election, it seemed best to present the brief to the privy council's consultative subcommittee on science policy, to which belonged many of the scientists consulted in preparation of the dossier.[36] In January 1958, the subcommittee informed CAP's directors that before the report went to cabinet, more precise propositions should be elaborated concerning site and management.

With the help of a second $5,000 grant from the NRC, a new committee was formed to study the question. A revised version of the report was presented to the executive committee in August 1958, and Kingston emerged as the best site, followed by Vancouver. The report suggested creation of a crown corporation, with the NRC as majority shareholder, although "other shares would be purchased by universities and other institutions in order to qualify their representatives as directors of the company."[37]

In the mean time, the Conservative government had been re-elected, and CAP was finally able to submit its report on 3 October 1958. During their meeting with the representative of the minister of industry, trade, and commerce (the minister having been held up in Europe by a conference), members of CAP's delegation were asked "questions concerning the attitude of university administrations to the project."[38] Now, as we have seen, at no time during preparation of the project did CAP inform the presidents of Canadian universities of their intentions. As the president of CAP B.W. Currie admitted: "While we could speak for physics departments generally, it was evident that we had nothing definite as far as University Presidents and Deans of Graduate Colleges were concerned."[39] Taken off-guard, Currie immediately wrote to the president of the National Conference of Canadian Universities (NCCU), which linked the presidents of all Canadian universities, to inform him of the project and to ask for the support of his executive.

It was a bit late and somewhat naive to believe that the NCCU would endorse a project of which it had not been informed. Having received a negative reply in December, the president of CAP wrote in desperation to each of the university presidents. "From the replies, Currie reported, that we received to our letter, it was evident that some believed that the large project would reduce, or at least limit, research funds distributed to the individual universities."[40] Meanwhile, the government itself had consulted a number of university presidents, and the minister's letter announcing the government's refusal to support the project noted: "It is felt that for the present, general support of physics in various fields in universities across the country is the most desirable way to spend whatever money is available."[41]

For having neglected the determinant role played by the presidents of Canadian universities in the development of the sciences and having wished to speak in their place, CAP failed in its attempt to establish a high-energy physics laboratory in Canada. Ten years later, in 1968, the federal government's decision to approve construction of a 500-MeV proton accelerator in Vancouver (TRIUMF) – a project implicating British Columbia's three universities as well as the University of Alberta – reminded CAP that even in physics, the presidents of Canadian universities had the last word.[42]

This episode represents a new stage in the formation of CAP's national identity, which could not be based merely on the geographical distribution of its members. As the existence of any group owes a good deal to representation work, in the double sense of showing and speaking for members' interests, a national association must find a national stage where it may meet and discuss with representatives of other groups.[43] The national high-energy physics laboratory project gave CAP its first occasion to have access to the national scene. Although the project itself did not work out, it was at least useful as a "laboratory exercise." As CAP's president, Currie, noted in his annual report to the general assembly in June 1959: "I think that the project was good for CAP. It gave us something to work for; it brought us to the attention of the Government, the Scientific Associations and the Universities as an organization prepared to speak with some authority on the needs for physics and physics research in Canada."[44]

This first experience occurred as Canada was preparing itself to enter what might be called the era of science policy. This was not by chance, for when science policy does not precede "Big Science," it is never very far behind. The 1963 Royal Commission on Government Organization (the Glassco Commission), and especially creation of the Science Secretariat in 1964 and of the Science Council in 1966, offered new opportunities for intervention to the different scientific societies, which would no longer need to "create" an event in order to make themselves seen on the national scene (as was the case with CAP).[45] Henceforth, they could content themselves (if they wished) with "reacting" to those events created by these new national organizations or, better still, collaborate with these organizations in preparing position papers to make "decision-makers" well aware of their claims.[46]

CONCLUSION

From a simple aggregate of individuals at the end of the First World War, Canadian physicists had been transformed by the 1950s into a well-constituted group. With its own official representatives, the group could now defend its long- and short-term interests in the constant negotiations between the many groups that oppose each other in Canadian society in order to define implicitly and explicitly the place due scientific research among other societal activities. The survival of a scientific discipline is in large part the result of the pressure the group can exert, directly or indirectly, particularly in elaborating discourses on the links between "pure" and "applied" science, research and industry, research and teaching, or even research and society. The ultimate purpose of these discourses is the survival of the group and the discipline to which it is devoted.

The participation of the academic fraction of Canadian physicists in the game and the stakes particular to the international field of physics thus depends to a great extent on these agents' ability to impose a representation of scientific activity which leaves as large a place as possible to a "disinterested" practice of physics, which can survive only as long as institutions "feed" it with grants, instruments, and especially students, the latter being the only ones able to assure the "future" of the group and its discipline.

Conclusion

The description of the efforts made by the first generation of researchers to modify the functioning of the institutions to which they were attached and to adapt them to their needs should not be read as the activities of a group that was well constituted from the very beginning and sought to impose its hegemony. On the contrary, the changes described here are an integral part of the formation of the group of Canadian physicists and of the corresponding discipline. Although this study has concentrated on a particular case, it seems clear that many scientific disciplines have emerged out of a similar process. In Canada, for example, chemistry and biology seem to have followed quite closely the path trod by physics. Moreover, these groups often spoke with a common voice on the development of university research and on modification of the rules of the Royal Society of Canada or of the mode of publication of its *Transactions.*

As Luc Boltanski has written, "The appearance of a new group is the product of a long-term structural rearrangement working on both objective properties and on representations."[1] In the case of Canadian physicists, this process began around 1850 and finished in 1950, as a mode of group representation determined in its essentials by the academic fraction of Canadian physicists.[2]

The emergence and subsequent institutionalization, within Canadian universities, of a new practice of physics, based on the production of new knowledge, were not the result of a single cause which worked over time to shape the image of the group of Canadian physicists. On the contrary, the transformation of the objective properties of agents and representations was the result of specific conjunctures which brought together many different developments.

In Canada, research in physics was a practice imported from Europe by J.J. Mackenzie and J.G. MacGregor at Dalhousie University in the late 1870s and, twenty years later, by H.L. Callendar at McGill. However, the presence of several researchers in Canadian universities was not sufficient to assure the constitution of the discipline of physics in Canada. Institutional conditions also had to be favourable to the exercise and development of this new practice of physics.

The first condition which made possible the appearance of the physics department as a discrete entity was the development of engineering education, itself the product of Canadian economic development in the 1870s. The "utilitarian" conception of science education which gave rise to the engineering programs also modified the traditional bachelor of arts program to make greater room for optional courses and advanced (honours) courses, thus introducing a certain level of specialization at the undergraduate level.

In my opinion, this transformation constitutes an important stage in the history of science in Canada. It furnished a basis for the institutional reproduction of future Canadian scientists. The appearance of "honours" sections in physics, for example, gave new visibility to physics as a discipline, which now corresponded to a career and was no longer a form of general knowledge that one acquired along with chemistry and Latin. Carried out during the 1870s and the 1880s, this restructuring of bachelor of arts studies served researchers who were beginning to appear in Canadian universities.

In the ten years prior to the First World War, the opening of universities in the Canadian west and the growth of universities such as Toronto and McGill led to the hiring of new physics professors, almost all of whom had been trained in research and who wished to continue to invest in this type of activity. These researchers' claims for better institutional conditions, favourable to the development of research, multiplied over the years but found few sympathetic ears until the First World War, which gave scientific research national stature by relating it to the country's industrial development.

Pressed by industry and the universities, the Canadian government created the National Research Council in 1916. Directed by those who had been the most ardent supporters of scientific research, the council soon established structures that favoured, above all, university research – which, according to NRC directors, would make possible development of industrial research. Thus, beginning in 1917, NRC scholarships for post-graduate study enabled more young Canadians to obtain master's and doctoral degrees in the main Canadian universities, while grants to researchers assured greater stability for university research, previously the recipient of little support.

These new structures ensured that physicists, chemists, and biologists could reproduce their discipline. For the first time, scientists could count on more regular recruitment of students to post-graduate education, where they acquired the researcher's "trade" and knowledge of the "rules of the game" which define the scientific field. Although master's and doctoral programs had been established at McGill and Toronto in 1906 and 1897, respectively, only creation of the NRC's scholarships made possible continuous production of graduates in the various sciences. With these changes, research became institutionalized toward the end of the First World War.

In the 1920s, rapid growth of scientific production and the effect of its institutionalization raised the problem of the diffusion of Canadian scientists' research results. Modification, begun in 1905, of the mode of publishing the Royal Society of Canada's *Transactions* – the only scientific journal of national stature – did not satisfy the needs of Canadian scientists. The concerted efforts of Canadian research scientists, physicists in particular, finally led the NRC to create, in 1929, the *Canadian Journal of Research*. The NRC thus solved a problem resulting from the growth in scientific production which was spurred on by the very research aid programs that the NRC had launched in 1917.

In order to assure the long-term reproduction of the discipline, physicists also tackled the problems of training and recruitment by looking into the teaching of physics in both secondary schools and universities. As no overproduction had been foreseeable in the 1920s, there had been little interest in industrial employment prospects, and, in general, physicists concentrated on problems typical of a discipline: recruitment of researchers, research budgets, publication of results, and the creation (failed) of a Canadian association of physicists.

In the 1930s, more graduates and the development of government and industrial research led to a greater array of positions available to physicists. Alongside the academic physicist appeared a new category, the industrial physicist, who had difficulty identifying with the image of the physicist projected by academics. The problems peculiar to this fraction of Canadian physicists became visible with the setting up, in 1945, of the Canadian Association of Professional Physicists (CAPP), whose objective was to procure for those physicists a status comparable to that of the engineer and to present a more pragmatic image of the physicist. Thus diversification of positions produced a crisis in physicists' mode of representation: employees of industry claimed professional status and control of the term "physicist"; academic physicists sought to endow themselves with the traditional tools of a discipline: meetings, journals, medals, and so on. Industrial physicists' efforts at professionalization failed, however, and the academics took charge of the organization under the name Canadian Association of Physicists (CAP).

At the start of the 1950s, CAP represented essentially the interests of academic physicists and was learning to represent them at the national level. The traditional image of the isolated physicist in the laboratory, preoccupied with the "laws of nature" – an image that industrial physicists opposed – had again become dominant and was used as a basis for recruiting new students.

In the 1960s, the increasing intervention of federal and provincial governments in the management of research caused the appearance of "science policy." In that sphere, CAP was able henceforth to enjoy its role as Canadian physicists' representative in front of "decision-makers," who, by determining the amounts to be allocated for research, control the primary resources which allow physicists, as well as other scientists, to continue to conduct "fundamental" and "disinterested" research. This increased dependence on public authority could not help but give greater importance to scientists' representatives, whose function is to uphold the belief in science, which is the ultimate but precarious foundation of university research.

Statistical Appendix

Figure A1
Physics Professors in Canadian Universities, 1900–48

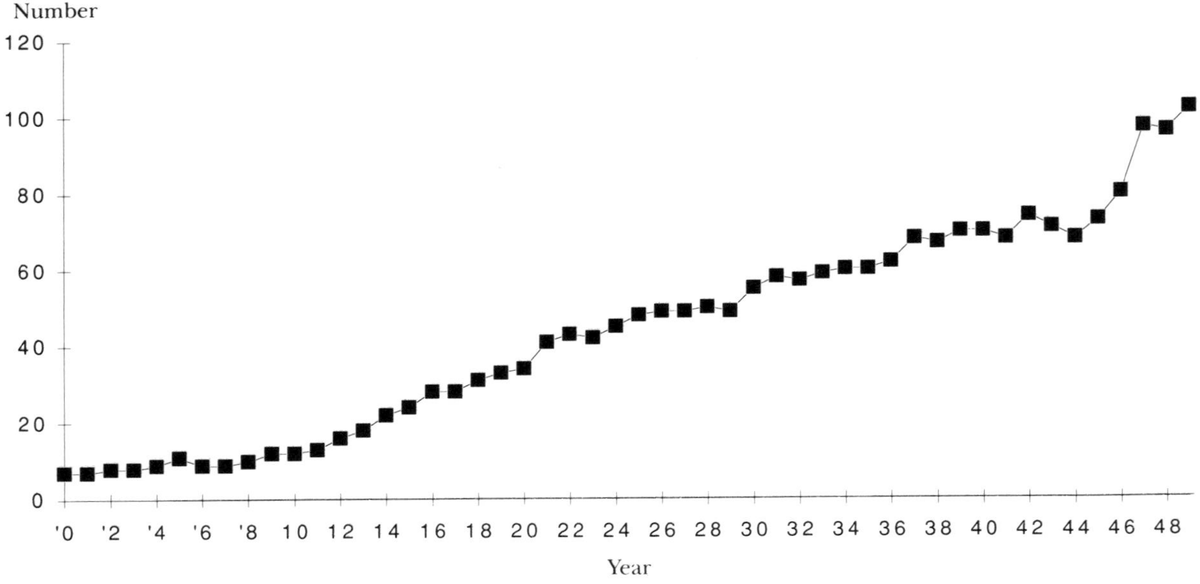

Figure A2
Numbers of MSCs and PHDs in Physics Given by Canadian Universities, 1900–60

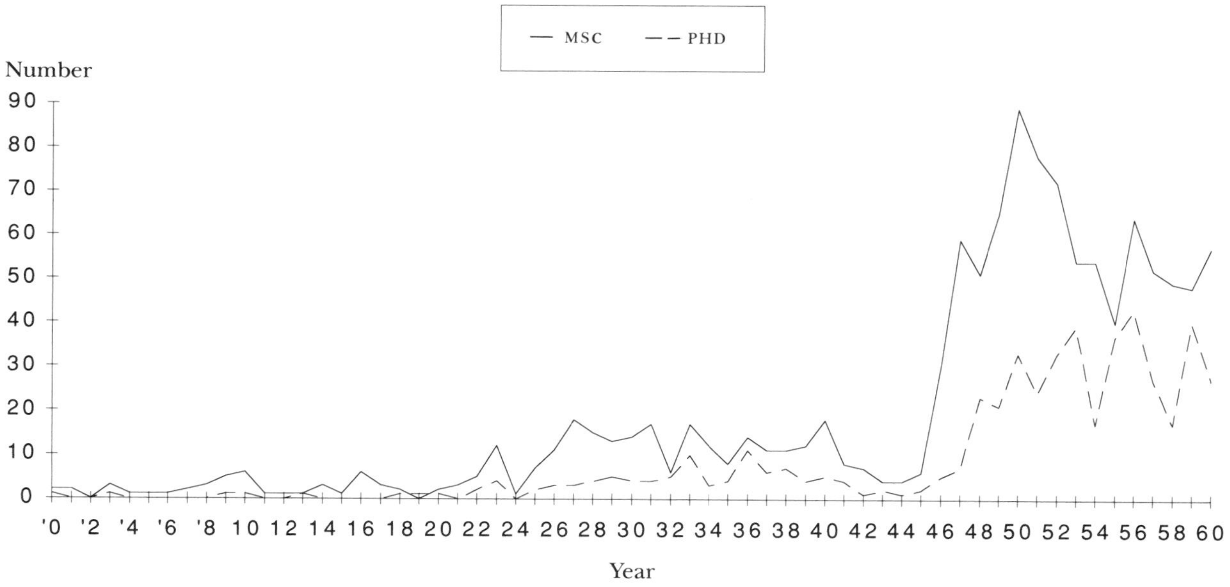

Figure A3
Publications in Physics in Canada, 1900–49

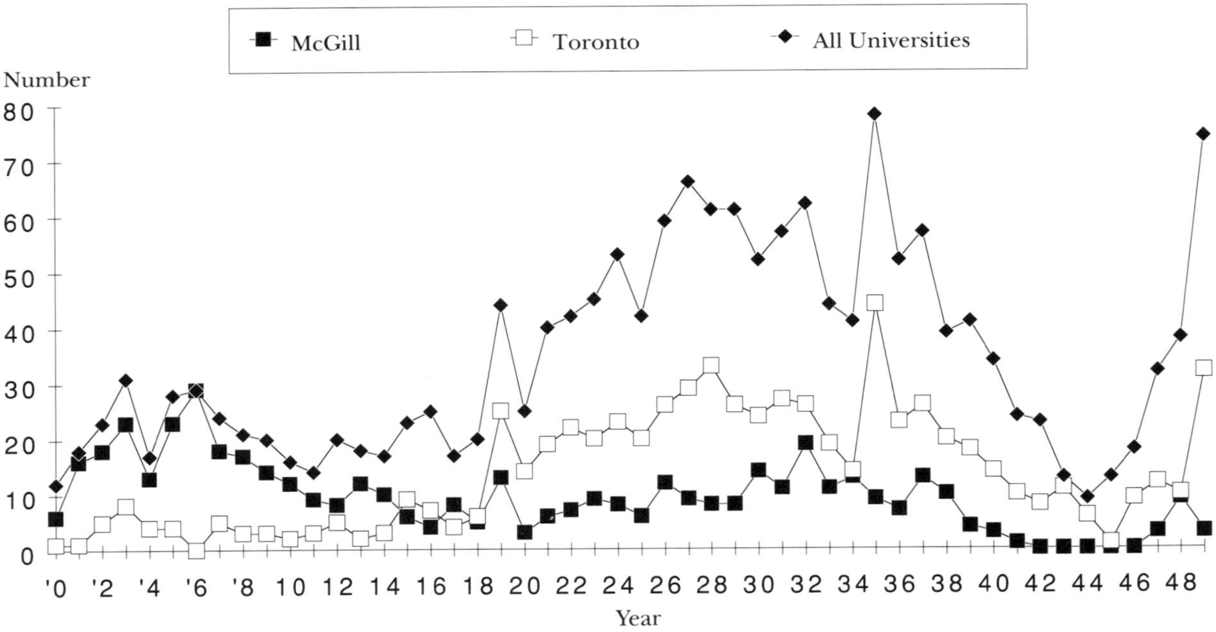

Table A4

Number of NRC Scholarships by Year and University (All Disciplines Included), 1917–37

Year	Toronto	McGill	Queen's	Alberta	Man.	Sask.	BC	Dalhousie	Montréal	Laval	UWO	McMaster
1917	3	3	–	1	–	2	–	–	–	–	–	–
1918	2	6	–	–	–	–	–	–	–	–	–	–
1919	9	6	2	–	–	1	1	–	–	–	–	3
1920	10	5	1	1	1	–	–	2	–	–	–	–
1921	21	11	2	–	1	3	2	1	–	–	–	–
1922	18	14	1	–	1	–	3	–	–	–	–	–
1923	16	13	5	2	3	1	3	–	–	–	–	–
1924	12	17	2	1	3	1	1	–	5	–	1	–
1925	17	14	3	1	1	2	–	1	2	–	1	–
1926	14	18	2	1	1	2	–	2	1	–	–	–
1927	16	21	3	1	1	1	–	3	–	–	1	–
1928	14	22	4	1	–	6	1	–	–	–	–	–
1929	13	28	1	1	–	4	–	1	1	–	–	–
1930	19	34	4	–	–	2	–	1	1	–	–	–
1931	15	27	4	–	–	2	–	1	–	1	–	–
1932	7	nd	5	nd	nd	nd	nd	nd	nd	nd	nd	nd
1933	6	8	1	–	–	1	–	–	–	1	–	–
1934	5	16	3	–	1	–	–	–	–	1	–	–
1935	4	20	2	–	–	1	–	1	–	1	–	–
1936	3	17	3	–	–	1	–	1	1	2	–	–
1937	9	24	4	–	–	1	–	1	1	1	1	–
Total	233	324	52	10	13	31	11	15	12	7	4	3

Source: NRC, Annual Reports.

Note: nd = no data.

Table A5
Number of NRC Scholarships by Discipline by Year (All Categories Included), 1917–37

Year	Physics	Chemistry	Life Sciences	Geology	Engineering	Mathematics
1917–18	5	4	–	–	–	–
1918–19	2	4	2	–	–	–
1919–20	4	10	7	1	–	–
1920–21	5	7	6	2	–	–
1921–22	11	16	10	3	1	–
1922–23	10	13	–	3	2	–
1923–24	11	19	6	7	–	–
1924–25	4	26	6	7	–	–
1925–26	12	25	4	1	–	–
1926–27	11	25	3	1	1	–
1927–28	18	19	7	3	–	–
1928–29	16	23	8	1	–	–
1929–30	16	26	4	1	1	1
1930–31	17	27	12	4	1	–
1931–32	11	28	6	3	1	1
1932–33	nd	nd	nd	nd	nd	nd
1933–34	7	6	4	–	–	–
1934–35	7	14	4	–	1	–
1935–36	12	13	4	–	–	–
1936–37	9	16	2	1	1	–
1937–38	10	24	5	1	1	1
Total	198	345	100	39	9	3

Source: NRC, Annual Reports.
Note: nd = no data.

Table A6
Number of Research Projects Funded by NRC by Year and by University, 1917–37

Year	Toronto	McGill	Queen's	Alberta	Man.	Sask.	BC	Dalhousie
1917–18	–	–	4	–	1	1	2	1
1918–19	–	3	–	–	–	1	–	–
1919–20	1	5	1	–	–	–	1	–
1920–21	1	2	2	1	1	–	1	–
1921–22	3	–	1	–	2	1	2	–
1922–23	3	3	2	1	2	–	1	–
1923–24	6	2	1	6	3	1	1	–
1924–25	3	1	1	3	2	1	–	–
1925–26	2	1	1	3	6	2	2	1
1926–27	4	5	–	6	4	2	2	1
1927–28	6	2	2	6	4	5	2	2
1928–29	3	2	2	6	4	5	2	–
1929–30	14	12	4	10	7	2	3	3
1930–31	10	9	6	12	10	8	4	2
1931–32	9	9	3	13	10	8	4	2
1932–33	nd	nd	nd	nd	nd	nd	nd	nd
1933–34	–	6	1	10	7	5	4	–
1934–35	1	4	1	6	3	5	1	–
1935–36	2	4	2	5	3	5	1	1
1936–37	3	5	2	1	3	1	1	–
1937–38	3	6	1	1	1	1	1	–
Total	74	81	37	90	73	54	35	13

Source: NRC, Annual Reports.
Note: nd = no data.

Table A7
Place of Publication of Canadian Physicists, 1900–39

	1900–17			1918–28			1929–39			
	Toronto	McGill	Other	Toronto	McGill	Other	Toronto	McGill	Other	NRC
Royal Society of Canada	30	56	6	119	26	40	48	7	3	0
CJR	0	0	0	0	0	0	41	17	40	47
Other	7	4	9	14	3	6	26	17	12	1
CANADA	37	60	15	133	29	46	115	41	55	48
Physical Review	13	14	19	1	3	23	16	6	28	0
American Journal of Science	4	5	2	0	0	1	0	0	0	0
Other	6	44	8	12	18	33	42	15	34	3
UNITED STATES	23	63	29	13	21	57	58	21	62	3
Philosophical Magazine	26	81	10	24	4	16	19	5	10	1
Transactions of the Royal Society of London	13	14	0	39	14	9	35	27	20	0
Nature	1	34	2	21	7	7	28	3	11	0
Other	6	1	1	4	1	6	8	4	10	0
GREAT BRITAIN	46	130	13	88	26	38	90	39	51	1

Table A7 (continued)

	1900–17			1918–28			1929–39			
	Toronto	McGill	Other	Toronto	McGill	Other	Toronto	McGill	Other	NRC
Zeit.F.Phys.	1	0	1	0	0	0	0	0	1	0
Phys.Zeit.	5	12	1	0	0	0	0	0	2	0
Other	0	2	2	0	0	0	9	0	1	0
GERMANY	6	14		0	0	0	9	0	4	0
FRANCE	1	1	0	1	0	0	1	1	1	0
OTHER COUNTRY	0	0	1	0	0	0	2	1	0	0
Total	113	268	60	235	76	141	275	103	173	52
		441			452			551		52

Notes

ABBREVIATIONS

BCAP   *Bulletin of the Canadian Association of Physicists*
CJR   *Canadian Journal of Research*
HSPS   *Historical Studies in the Physical Sciences*
MSRC   *Mémoires de la Société royale du Canada*
NAC   National Archives of Canada, Ottawa
PNRC   *Proceedings of the National Research Council*
PRSC   *Proceedings*, Royal Society of Canada
TRSC   *Transactions*, Royal Society of Canada

INTRODUCTION

1 R.S. Turner, "The Growth of Professional Research in Prussia, 1818 to 1848. Causes and Context," *HSPS*, 3 (1971), 137–82; Karl Hufbauer, *The Formation of the German Chemical Community* (Berkeley 1982); Terry Shinn, "The French Science Faculty System, 1808–1914: Institutional Change and Research Potential in Mathematics and the Physical Sciences," *HSPS*, 10 (1979), 271–332; Craig Zwerling, "The Emergence of the Ecole Normale Supérieure as a Center of Scientific Education in the Nineteenth Century," in R. Fox and G. Weisz, eds., *The Organization of Science and Technology in France, 1808–1914* (Cambridge 1980), 31–60; Dominique Pestre, *Physique et physiciens en France 1918–1940* (Paris 1984); R. Sviedrys, "The Rise of Physical Science in Victorian Cambridge," *HSPS*, 2 (1970), 127–45, and "The Rise of Physical Laboratories in Britain," ibid., 7 (1976), 405–36.

For the United States, see N. Reingold, ed., *The Sciences in the American Context: New Perspectives* (Washington 1979), and A. Oleson and J. Voss, eds., *The Organization of Knowledge in Modern America* (Baltimore 1979);

Roger L. Geiger, *To Advance Knowledge: The Growth of American Research Universities* (New York 1986). For an overview, see D. Kevles, *The Physicists: The History of a Scientific Community in Modern America* (New York 1978). For Japan, see K. Koizumi, "The Emergence of Japan's First Physicists: 1868–1900," *HSPS*, 6 (1975), 1–107, and for Australia, see R.H. Home, "Origins of the Australian Physics Community," *Historical Journal*, 20 (1982–83), 383–400.

2 In spite of a wealth of documentation, Kevles's book on the history of physics in the United States is not constructed in relation to a precise problematic. It would therefore be difficult to use this book for a comparative study. By describing a series of events as if their order were to be taken for granted, the book answers few of those questions that might be asked by a historian or sociologist interested in the mechanisms of the formation of a scientific community. In contrast, Hufbauer's book is structured according to a plan which clearly shows the different stages in the formation of the German community of chemists at the end of the eighteenth century. See Kevles, *The Physicists*, and Hufbauer, *The Formation*.

3 This conception of an educational system has been developed by Pierre Bourdieu and Luc Boltanski in "The Educational System and the Economy: Titles and Jobs," in Charles C. Lemert, ed., *French Sociology: Rupture and Renewal since 1968* (New York 1981), 141–51. See also P. Bourdieu and J.C. Passeron, *Reproduction* (London, 1977).

4 On the concept of habitus, see P. Bourdieu, *Outline of a Theory of Practice* (Cambridge 1977) and "Systems of Education and Systems of Thought," *Social Science Information*, 14 no. 3 (1967), 338–58.

5 As Emile Durkheim has written: "Man does not change arbitrarily; he does not metamorphose at will at the sound of an inspired prophet's voice, for all transformation having to confront an organized and inherited past, is hard and laborious. It is only accomplished under the reign of necessity." *L'évolution pédagogique en France* (Paris 1969), 377.

6 W.O. Hagstrom, *The Scientific Community* (New York 1975), 9.

7 Pierre Bourdieu, "The Specificity of the Scientific Field and the Social Conditions of the Progress of Reason," in Lemert, *French Sociology*, 257–92.

8 Pierre Bourdieu, *Questions de sociologie* (Paris 1980), 114.

9 J.B. Morrell, "The Chemists' Breeders: The Research Schools of Liebig and Thomas Thomson," *Ambix*, 19 (1972), 1–46; B.W.J. Olt, "Social Aspects in the Emergence of Chemistry as an Exact Science: The British Chemical Profession," *British Journal of Sociology* no. 21 (1970), 181–99; Kenneth L. Caneva, "From Galvanism to Electrodynamics: The Transformation of German Physics and Its Social Context," *HSPS*, 9 (1978), 137–8. For a magisterial study of the emergence and institutionalization of physical research in German universities, see

Christa Jungnickel and Russell McCormach, *Intellectual Mastery of Nature: Theoretical Physics from Ohm to Einstein* (Chicago 1986), 2 vols.

10 Although I am using this model only to analyse the Canadian situation, I believe that it has a much more general bearing and that it is applicable to countries such as Japan and Australia which, like Canada, imported their conception of scientific research from Europe. Koizumi's study, for example, describes the emergence of research in a manner quite similar to that which we have observed in Canada, although he does not treat the institutionalization of this activity. Similarly, using the category of professionalization, Home in reality analysed the processus of the definition of the social identity of Australian physicists. See R.H. Home, "Between Classroom and Industrial Laboratory: The Emergence of Physics as a Profession in Australia," *The Australian Physicist*, xx (Aug. 1983), 163–7. Other university disciplines can be studied using the model here advanced. See, for example, Thomas L. Haskell, *The Emergence of Professional Social Science* (Urbana, Ill., 1977); William R. Keylor, *Academy and Community: The Foundation of the French Historical Profession* (Cambridge 1975); and Gerald Graff, *Professing Literature: An Institutional History* (Chicago 1987).

11 A similar conception of narrative is defended by Martin Rudwick in "The Revival of Narrative" in his book *The Great Devonian Controversy* (Chicago 1985), 11–14.

CHAPTER ONE

1 For further detail on the birth of Canadian universities and colleges during the first half of the nineteenth century see Robin S. Harris, *A History of Higher Education in Canada* (Toronto 1976), chap. 1. For the development of railways in Canada, see H.T. Easterbrook and H.G.J. Aitken, *Canadian Economic History* (Toronto 1958), chap. 14, and G.P. de T. Glazebrook, *A History of Transportation in Canada*, I (Toronto 1964), chap. 5.

2 Stanley B. Frost, *McGill University: for the Advancement of Learning* Volume 1, 1801–1895 (Montreal 1980), 185.

3 Ibid., 176.

4 Ibid., 182. See also John William Dawson, *Fifty Years of Work in Canada* (London 1901), 70–9.

5 Cited in A. Foster Beard, "The History of Engineering at the University of New Brunswick," in *The University of New Brunswick Memorial Volume* (Fredericton 1950), 76.

6 Frost, *McGill University*, I, 181–4; Dawson, *Fifty Years*, 87. For more detail on the history of science education in New Brunswick, see Richard A. Jarrell, "Science Education at the University of New Brunswick in the Nineteenth Century," *Acadiensis* (spring 1973), 55–79.

7 The management of the institution includes not only administrative organization but also the tradition of intramural residence more suited to the training of future clergymen than to that of citizens expected to learn how to live in society. In this respect, Dawson wrote in his contribution to the report: "The residence of pupils within the College buildings is not of such utility as has hitherto been supposed ... I cannot doubt that College residence is, even under the most favourable circumstances, more dangerous to the health, manners and morals of the students than to reside in respectable private houses." Cited in Frost, *McGill University*, I, 183.

8 Ibid., 185.

9 Ibid., 187–8.

10 C.R. Young, *Early Engineering Education at Toronto 1851–1919* (Toronto 1958).

11 At the University of New Brunswick "the number of students receiving diplomas in Engineering [was] not more than five or six yearly, usually less than this number"; Beard, "The History of Engineering," 81. For Toronto, see Harris, *History of Higher Education*, 164, and Young, *Early Engineering Education*, 18.

12 See Luc Chartrand, Raymond Duchesne, and Yves Gingras, *Histoire des sciences au Québec* (Montreal 1987), 227–33; Yves Gingras and Robert Gagnon, "Engineering Education and Research in Montreal: Social Constraints and Opportunities," *Minerva*, 26 (spring 1988), 53–65.

13 From 1870 to 1890, the primary and secondary industrial sectors grew at a mean annual rate of 4.6 per cent, which is comparable only with the post–Second World War "boom." This growth slowed during the 1890s. See G.W. Bertram, "Economic Growth in Canadian Industry 1870–1915," in W.T. Easterbrook and M.H. Watkins, eds., *Approaches to Canadian Economic History* (Toronto 1967), 75–98. For the history of civil engineers in Canada, see J. Rodney Millard, *The Master Spirit of the Age: Canadian Engineers and the Politics of Professionalism* (Toronto 1988).

14 Cited by Young, *Early Engineering Education*, 24.

15 For the regional aspect of economic development, see Bertram, "Economic Growth," 96–8 and Easterbrook and Aitken, *Canadian Economic History*, 250.

16 H. Neatby, *Queen's University: I, 1841–1917: And Not to Yield* (Montreal 1978).

17 Harris, *History of Higher Education*, 145.

18 James Loudon, "The Evolution of the Physical Laboratory," *University of Toronto Monthly*, 8 (1907), 43.

19 Ibid., 43.

20 Ibid., 44.

21 Harris, *History of Higher Education*, 145.

22 Hugh Hawkins, "University Identity: The Teaching and Research Functions," in Alexandra Oleson and John Voss, eds., *The Organization of Knowledge in Modern America 1860–1920* (Baltimore 1979), 293.

23 Frost, *McGill University*, I, 273.

24 Harris, *History of Higher Education*, 217.

25 Quoted by Beard, "The History of Engineering," 82.

26 *Formal Opening of the Engineering and Physics Building McGill University, Montreal Feb. 24th 1893.* McGill University Archives (MUA) Acc. 409, 35/3/1 a, 40.

27 In a letter to the minister of education opposing the separation of the School of Applied Sciences, Loudon wrote: "Even if the proposed separation be regarded as an expedient to relieve the congestion of the present laboratories, it must prove itself ineffective owing to the fact that increased accommodation is already required for the proper teaching of the students in arts, and in the interests of the University cannot longer be delayed. In a word, the number of students in Arts have so increased that the accommodation already asked for in the proposed Physics building is urgently required, entirely irrespective of the Engineering students." Loudon to Harcourt, 13 Sept. 1904, University of Toronto Archives (UTA), Loudon Collection, Box 4.

28 *Victoria University Callendar*, 1862–63, quoted by Harris, *History of Higher Education*, 131.

29 W.J. Alexander, ed., *The University of Toronto and Its College 1827–1906* (Toronto 1906), 99.

30 Harris, *History of Higher Education*, 40–2.

31 For more on these transformations, see Patricia Jane Jasen, "The English Canadian Liberal Arts Curriculum: An Intellectual History, 1800–1950," PHD thesis, University of Manitoba, 1987, chap. 2.

32 Harris, *History of Higher Education*, 40–5.

33 Elizabeth J. Allin, *Physics at the University of Toronto 1843–1980* (Toronto 1981), 4.

34 For a detailed study of these transformations, see Michael Sanderson, *The Universities and British Industry 1850–1970* (London 1972).

35 Ibid., 148.

36 For further information on the transplantation of English and Scottish models to Canada, see Harris, *History of Higher Education*, chap. 3.

37 Jarrell, "Science Education at the University of New Brunswick," 58–61 and 72–3.

38 Neatby, *Queen's University*, 58.

39 Peter J. Bowler, "The Early History of Scientific Societies in Canada," in A. Oleson and S. Brown, eds., *The Pursuit of Knowledge in the Early American Republic* (Baltimore 1976), 326–39.

40 For Canadian interest in meteorology, see Suzanne Zeller, *Inventing Canada* (Toronto 1987).

41 John A. Patterson, "John Bradford Cherriman: An Appreciation,"
    University Monthly, 9 (1908), 77–83.
42 For further detail on the history of the observatory, see Richard A. Jarrell
    "Origins of Canadian Government Astronomy", *Journal of the Royal
    Astronomical Society of Canada*, 69 (1975), 77–85. For astronomy in Canada
    see also, by the same author, *The Cold Light of Dawn: A History of Astronomy
    in Canada* (Toronto 1987). With regard to the "mathematical tripos" and
    the scientific training of Cambridge graduates, see David B. Wilson,
    "Experimentalists among the Mathematicians: Physics in the Cambridge
    Natural Sciences Tripos, 1851–1900," *HSPS*, 12, no. 2 (1982), 325–71,
    and Harvey W. Becher, "William Whewell and Cambridge Mathematics,"
    HSPS, 11, no. 1 (1980), 1–48.
43 *TRSC*, 1 (1882), section III, 15.
44 H.H. Langton, *James Loudon and the University of Toronto* (Toronto
    1927).
45 *Canadian Journal*, 14 (second series), 1875, 62.
46 Ibid., 354, and 13 (1873), 231.
47 "Alexander Johnson," *PRSC*, 7, (third series), 1913, xii.
48 A. Johnson, "The Preparation in Montreal," *TRSC*, 1 (1883),
    section III, 83
49 A. Johnson, "On the Need of a 'Coast Survey' for the Dominion of
    Canada," *TRSC*, 11 (1893), section III, 55. For further detail on these
    events, see Vittorio G.M. de Vecchi, "The Dawning of a National
    Scientific Community in Canada, 1878–1896," *Scientia Canadensis*, 8
    no. 1 (June 1984), 47–8.
50 A. Johnson, "Newton's Use of Slit and Lens in Forming a Pure
    Spectrum," *TRSC*, 9 (1891), section III, 49
51 Upon his death, it was written: "He had not the practical training
    requisite for research in experimental physics but devoted himself to the
    teaching and administrative side." *Nature*, 6, no. 3 (1923), 817.
52 John Cox, "Apparently Accidental Formation of Frazil Ice in a
    Cryophorous," *TRSC* (second series), 10 (1904), section III, 4. On Cox's
    relationship with the discovery of x-rays, see Yves Gingras, "La réception
    des rayons x au Québec : radiographie des pratiques scientifiques," in
    Othmar Keel, Marcel Fournier, and Yves Gingras, eds., *Sciences et médecine
    au Québec: perspectives sociohistoriques* (Québec 1986), 69–86.
53 On this definition of a generation, see Pierre Bourdieu, *Distinction*
    (Cambridge 1984).
54 On the notion of habitus as a system of schemes that generates practices,
    perceptions, and evaluations of practices, see P. Bourdieu, *Outline of
    a Theory of Practice* (Cambridge 1977). See also, by the same author,
    "Systems of Education and Systems of Thought," *Social Science
    Information*, 14 (1967), 338–58.
55 J.B. Cherriman, *Mechanics* (Toronto 1858); J. Cox, *Mechanics* (Cambridge

1904); J. Loudon, *Algebra* (Toronto 1873) and *Algebra for Beginners* (Toronto 1876).

56 For the history of the classical colleges in Quebec and their relationship with Laval University, see Claude Galarneau, *Les collèges classiques au Canada français* (Montreal 1978).

57 Frost, *McGill University*, I, 172–4.

58 Unfortunately, there are still no institutional histories of Quebec francophone universities comparable to those concerning McGill, Toronto, and most other anglophone universities in Canada. The only source remains Honorius Provost, *Historique de la Faculté des Arts de l'Université Laval, 1852–1952* (Quebec 1952).

59 "Thomas-Etienne Hamel," *MSRC*, 8 (1914), v.

60 "Joseph Clovis Kemner Laflamme," ibid. (2e série), 5 (1911), v.

61 For more detail on the origins of the Royal Society of Canada, see Vitorrio M.G. de Vecchi, "Science and Government in Nineteenth-Century Canada." PHD thesis, University of Toronto, 1978, and "Dawning," 32–58. The origins of the Royal Society of Canada are also discussed in Peter J. Bowler, "The Early History of Scientific Societies in Canada," in A. Oleson and S. Brown, eds., *The Pursuit of Knowledge in the Early American Republic* (Baltimore 1976), 326–39; R. Daley and P. Dufour, "Creating a "Northern Minerva': John William Dawson and the Royal Society of Canada," *HSTC Bulletin*, 5 no. 1 (Jan. 1981), 3–14, and Richard A. Jarrell, "The Influence of Irish Institutions upon the Organization and Diffusion of Science in Victorian Canada," *Scientia Canadensis*, 9 no. 2 (1985), 150–64.

62 On Laflamme see Raymond Duchesne, "Science et société coloniale : les naturalistes du Canada-français et leurs correspondants scientifiques (1860–1900)," *HSTC Bulletin*, 18 (May 1981), 99–139.

63 Ibid., 116.

64 *MSRC*, 3 (1884), 91–100, and 10 (1891), 3–7. On Hamel and Laflamme's positions with regard to Darwinism, see Chartrand, Duchesne, and Gingras, *Histoire des sciences au Québec*, chap. 6.

65 Duchesne, "Science et société coloniale," studies the details of the itineraries of these two botanists. See also, by the same author, "La bibliothèque scientifique de l'abbé Provancher," *Revue d'histoire de l'Amérique française* 34 (March 1981), 535–56.

66 Of course, classification requires a certain amount of training. Provancher, for example, had to be corrected in his classification of the hymenopters. See Chartrand, Duchesne, and Gingras, *Histoire des sciences au Québec*, 190–1.

67 Roy Porter, "Gentlemen and Geology: The Emergence of a Scientific Career, 1660–1920," *Historical Journal*, 21 no. 4 (1978), 809–36.

68 "Henri Simard," *MSRC* (third series), 22 (1928), iv.

69 For further detail on the development of the sciences in francophone

Quebec after 1920, see Cyrias Ouellet, *La vie des sciences au Canada français* (Quebec 1964); Raymond Duchesne, *La science et le pouvoir au Québec (1920–1965)* (Quebec 1978), and Chartrand, Duchesne, and Gingras, *Histoire des sciences au Québec,* chaps. 8, 11, 12.

70 Roy M. Macleod, "The Support of Victorian Science: The Endowment of Research Movement in Great Britain, 1868–1900," *Minerva,* 9 no. 2 (1971), 197–230.

71 For further detail on the emergence of physics laboratories, see R. Sviedrys, "The Rise of Physics Laboratories in Britain," *HSPS,* 7 (1976), 405–36, and "The Rise of Physical Science at Victorian Cambridge," *HSPS,* 2 (1970), 127–45.

72 J.B. Morrell, "The Chemists' Breeders: The Research Schools of Liebig and Thomas Thomson," *Ambix,* 19 (1972), citation on p. 2.

73 J.C. Poggendorff, *Biographisch-Literarisches Handwörterburch zur Geschichte der Exackten Wissenschaften,* I (Amsterdam 1965), 852.

74 Quoted by J.H.L. Johnstone, *A Short History of the Physics Department, Dalhousie University, 1838–1956* (Halifax 1971), 8.

75 Minutes of the Board of Governors, Dalhousie University Archives (DUA), vol. 3, 14 Sept. 1877.

76 Ibid., 21 Aug. 1879.

77 "James Gordon MacGregor," *Proceedings of the Royal Society of London,* 89, series A (1913), xxvi.

78 Robert A. Falconer, "The Gilchrist Scholarship: An Episode in the Higher Education of Canada," *TRSC* (third series), 27 (1933), section II, 5–13.

79 J.G. MacGregor and J.A. Ewing, "The Conductivity of Certain Saline Solutions, with a Note on the Density," *Transactions of the Royal Society of Edinburgh,* 27 (1873), 51–70.

80 For further detail on this period, see John W. Servos, "Physical Chemistry in America, 1890–1933: Origins, Growth, and Definition," PHD thesis, Johns Hopkins University, 1979, chap. 1.

81 *TRSC,* 1 (1882), section III, 21.

82 See, for example, J.G. MacGregor, "On Carnot's cycle in Thermodynamics," *Transactions of the Nova Scotia Institute of Science,* 7 (1888–89), 227.

83 *Application of James Gordon MacGregor for the Professorship of Natural Philosophy in the Edinburgh University,* 1901, 2–3, Public Archives of Nova Scotia, MG 100, vol. 182, no. 37.

84 Ibid., iv–v.

85 P. Forman, J.L. Heilbron, and S. Weart, *Physics circa 1900: Personnel, Funding and Productivity of the Academic Establishment, HSPS,* 5 (Princeton 1975), 92.

86 Frost, *McGill University,* I, 43, 273–4.

87 J.J. Thomson, *Recollections and Reflections* (New York 1937), 435; my emphasis.
88 John L. Heilbron, "Physics at McGill in Rutherford's Time," in M. Bunge and W.R. Shea, eds., *Rutherford and Physics at the Turn of the Century* (New York 1979), 47. See also Lewis Pyenson, "The Incomplete Transmission of a European Image: Physics at Greater Buenos Aires and Montreal," *Proceedings of the American Philosophical Society*, 122 no. 2 (1978), 99–114.
89 *Annual Report of the President*, McGill University, 1894, 17.
90 *Annual Report*, McGill University, 1895, 24.
91 Pyenson, "Incomplete Transmission," 106.
92 H.A.M. Snelders, "Callendar, Hugh Longbourne," *Dictionary of Scientific Biography*, IV (New York 1971), 19–20.
93 I cannot here discuss the content of Rutherford's work, which has been the subject of a large number of publications by historians of physics. See, for example, A.S. Eve, *Rutherford* (London 1939); Thaddeus J. Trenn, *The Self-Splitting of the Atom: The History of the Rutherford-Soddy Collaboration* (London 1977) and "Rutherford in the McGill Physical Laboratory," in M. Bunge and W. Shea, eds., *Rutherford and Physics*, 89–109; David Wilson, *Rutherford: Simple Genius* (Cambridge 1983).
94 Rutherford to Mary Newton, 30 July 1898, in Eve, *Rutherford*, 54–7, 64; Heilbron, "Physics at McGill," 43.
95 Quoted by H.H. Langton, *Sir John Cunningham McLennan: A Memoir* (Toronto 1939), p. 15.
96 Allin, *Physics at the University of Toronto*, 7–23.
97 McGill thus joined other English-Canadian universities which had adopted this diploma – distinct from yet equivalent to the traditional BA – at the beginning of the 1890s. At the time, Laval also offered a BSC, but it was worth less than the BA. See Harris, *History of Higher Education*, 126.
98 The onset of the First World War allowed King to meet up with his mentor in England at the Imperial College of Science and Technology. The biographical information in this section comes from different sources: *American Men of Science: List of Graduates of the Faculty of Applied Science, McGill University* (1913); and *Report of the Commissioners for the Exhibition of 1851* (London 1930).
99 Rutherford visited her while passing through Montreal in 1914. She then had three children – "all admirable specimens" – Rutherford wrote in a letter to his wife; Eve, *Rutherford*, 231. On her stay at Bryn Mawr, see M. Rossiter, *Women Scientists in America: Struggles and Strategies to 1940* (Baltimore 1982), 15–16. For more detail on her career, see M.F. Rayner-Canham and G.W. Rayner-Canham, "Harriet Brooks, 1876–1933: Canada's First Woman Nuclear Physicist," in Marianne Gosztonyi-Ainley, ed., *Despite the Odds: Essays on Canadian Women and Science* (Toronto 1989), 185–203.

100 In the catalogue of Rutherford's correspondence, McClung's letters are spread out between 1901 and 1905, Cooke's between 1903 and 1918, and Boyle's between 1907 and 1936. L. Badash, *Rutherford Correspondence Catalogue*, American Institute of Physics, (New York 1974).

101 Servos, "Physical Chemistry in America," 497.

102 F. Spitzer and E. Silverster, eds., *McGill University Thesis Directory, 1881–1959* (Montreal 1976).

CHAPTER TWO

1   I will use the term "National Research Council of Canada" or "NRC" even though this name did not become official until 1925. Before this date, the NRC was known as the Honorary Advisory Council for Scientific and Industrial Research. In practice, however, from the beginning, the institution was referred to as the "Research Council." See M.L. Thistle, *The Inner Ring: The Early History of the National Research Council* (Toronto 1966), 131.

2   Hugh Hawkins, "University Identity: The Teaching and Research Functions," in Alexandra Oleson and John Voss, eds., *The Organization of Knowledge in Modern America, 1860–1920* (Baltimore 1979), 25.

3   Robert A. Falconer, "The Gilchrist Scholarship: An Episode in the Higher Education of Canada," *TRSC* (third series), 27 (1933), section II, 5–13.

4   Minutes of the Faculty of Arts, 10 Dec. 1886, Dalhousie University Archives (DUA).

5   Falconer, "The Gilchrist Scholarship," 10.

6   "William Lawton Goodwin (1856–1914)," *TRSC* (third series), 39 (1945), 87–8; "James Gordon MacGregor, 1852–1913," *Proceedings of the Royal Society of London* (series A, 89 (1913); xxvi–xxviii; on S.W. Hunton, see John G. Reid, *Mount Allison University, Vol. I, 1843–1914* (Toronto 1984), 185–6.

7   Figures from P.N. Ross, "The Development of the PHD at the University of Toronto, 1871–1932," EDD thesis, University of Toronto, 1972, 181. For the 1890s, Robin Harris gives the following statistics: more than thirty at Johns Hopkins, more than sixty at Harvard, more than fifty at Cornell, and more than eighty at Chicago. R.S. Harris, *The History of Higher Education in Canada* (Toronto 1976), 191–2.

8   Ross, "Development of the PHD," 199; Harris, *History of Higher Education*, 188.

9   Ross, "Development of the PHD," 89.

10 P.N. Ross, "The Establishment of the PHD at Toronto: A Case of American Influence," in M.B. Katz and P.H. Mattingly, eds., *Education and Social Change* (New York 1973), 193–214.

11 Ibid., p. 162. On Macallum see the *Dictionary of Scientific Biography* (New York 1973), VIII, 583–4; *PRSC* (third series), 28 (1934), xix–xxi; and Sandra F. Mcrae, "A.B. Macallum and Physiology at the University of Toronto," in Gerald L. Geison, ed., *Physiology in the American Context: 1850–1940* (Bethesda 1987), 97–114.

12 J. Loudon, "Changes and Progress," *University of Toronto Monthly*, 1 (July 1897), 6–9; cited in Ross, "Development of the PHD," 118.

13 Ross, "Development of the PHD," 118–96.

14 A.B. Macallum, "The Foundation of the Board of Graduate Studies," *University of Toronto Monthly* 16 (1916), 220. On Macallum, see *Dictionary of Scientific Biography*, VIII (New York 1973), 583–4; also *TRSC* (third series), 28 (1934), xix–xxi.

15 Stanley B. Frost, *McGill University: For the Advancement of Learning, Vol. II, 1895–1971* (Montreal 1984), 80–2.

16 Ibid., 177.

17 Ross, "Development of the PHD," 292.

18 Ibid., 245. For further discussion of the AAU, see L. Veysey, *The Emergence of the American University* (Chicago 1965), 175–7.

19 Loudon, "The Universities in Relation to Research," *TRSC* (second series), (1902) lvii.

20 Ibid., lvi.

21 Frost, *McGill University*, II, 177.

22 Ross, "Development of the PHD," 195. Until the end of the Second World War, these two institutions were the only ones to offer doctoral programs in most disciplines. See W.P. Thompson, *Graduate Education in the Sciences in Canada* (Toronto 1963).

23 Roy MacLeod, "The Endowment of Research Movement in Great Britain, 1868–1900," *Minerva* 9 (1971), 197–230.

24 Roy M. MacLeod and E. Kay Andrews, "Scientific Careers of 1851 Exhibition Scholars," *Nature* 218 (15 June 1968), 1,012. On the financial aid program of the École pratique, see Craig Zwerling, "The Emergence of the École Normale Supérieure as a Centre of Scientific Education in the Nineteenth Century," in R. Fox and G. Weisz, ed., *The Organization of Science and Technology in France, 1808–1914* (Cambridge 1980), 45.

25 *Record of the Science Research Scholars of the Exhibition of 1851* (London 1930), 1.

26 Minutes of the Board of Governors, vol. 4, 8 Nov. 1892, p. 226, Dalhousie University Archives (DUA), MS IS 4.

27 Ibid., 227.

28 Ibid.

29 The fourth conference of Canadian universities, in 1917, resolved to send a circular letter to the major British universities arguing that "[only with] the establishment of doctorates that may be obtained within a

reasonable time ... can we hope that the stream of students which of late has set towards the United States, will be directed to the Universities of Britain." See Fourth Conference of Canadian Universities, 24–25 May 1917, in *National Conference of Canadian Universities* (n.p., n.d.), 63.

30 Harris, *History of Higher Education*, 315.

31 On the importance of these means of production as a factor in the accumulation of symbolic capital, one may profitably read Bruno Latour and Steve Woolgar, *Laboratory Life: The Social Construction of Scientific Facts* (London 1979), chap. 5.

32 W.L. Morton, *One University: A History of the University of Manitoba 1877–1952* (Toronto 1957).

33 Ibid., 62.

34 At McGill, the PHD was awarded starting in 1906. It replaced the DSC, which was given as an honorary award to those who had published several papers. See Frost, *McGill University*, II, 80–2.

35 Charles M. Johnston, *McMaster University, Vol. I, The Toronto Years* (Toronto 1970), 109–16.

36 W.P. Thomson, *The University of Saskatchewan: A Personnel History* (Toronto 1970), 121–2.

37 Harry T. Logan, *Tuum Est: A History of the University of British Columbia* (Vancouver 1958).

38 Ibid., 25–51.

39 A.L. Clark, *The First Fifty Years: A History of the Science Faculty at Queen's University 1893–1943* (Kingston 1944).

40 I have included among physicists considered Canadian three who were born abroad but who received their university education in Canada: place of education, not place of birth, determines career choices. For example, A.N. Shaw was born in England but received his BA from McGill, like his colleague H.T. Barnes, who was born in the United States. The biographical data is taken from successive editions of *American Men of Science* (New York).

41 Ross, "Development of the PHD," 297.

42 K.R. Sopka, *Quantum Physics in America* (New York 1980), A–31.

43 R. Home, "Origins of the Australian Physics Community," *Historical Studies*, 20 (1983), 386.

44 *Application of James MacGregor, Monro Professor of Physics, to the Professorship of Natural Philosophy in the Edinburgh University* (Halifax 1901), 2–3 MG 100, vol. 182, no. 37, Public Archives of Nova Scotia (PANS).

45 *PRSC* (second series), 4 (1898), v.

46 Ibid.

47 *Annual Report of the Principal: McGill University*, 1892–3, 28.

48 H.A.M. Snelders, "Hugh Longbourne Callendar," *Dictionary of Scientific Biography*, III (New York 1973), 19–21. See also John L. Heilbron, "Physics at McGill in Rutherford's Time," in M. Bunge and W. Shea, eds.,

Rutherford and Physics at the Turn of the Century (New York 1979), 42–73.

49 *Annual Report of the Principal: McGill University*, 1895–6, 28.

50 Ibid., 1900–1, 32.

51 Ibid., 1901–2, 27; 1902–3, 35. See also Arthur S. Eve, *Rutherford* (London 1939), 86–9, 92, and Lawrence Badash, "The Origins of Big Science: Rutherford at McGill," in Bunge and Shea, eds., *Rutherford and Physics*, 23–41.

52 *Annual Report of the Principal: McGill University*, 1902–3, 35.

53 *Annual Report of the President, McGill University*, 1903–4, 40.

54 *Annual Report of the Principal, McGill University*, 1905–6, 49.

55 *Annual Report of the President, Queen's University*, 1911–12, 15.

56 *Annual Report of the President, McGill University*, 1913–14, 71. *PRSC* (third series), 9 (1915), xliv. King received $300 from the Royal Society for his research on the influence of atmospheric conditions on the propagation of sound waves.

57 *Annual Report of the President, McGill University*, 1909–10, 67.

58 Ibid., 1910–11, 68.

59 E.F. Burton to J.C. McLennan, 29 Sept. 1908, University of Toronto Archives (UTA), Falconer Papers, Box 6.

60 J.C. McLennan to J. Loudon, 7 Sept. 1897, UTA, Loudon Papers, Box 7.

61 J.C. McLennan to R. Falconer, 29 Sept. 1908, UTA, Falconer Papers, Box 6.

62 *Annual Report of the President, Queen's University*, 1901–2, 95.

63 Ibid., 1911–12, 15.

64 W.C. Baker, "The Cavendish Laboratory," *Queen's Quarterly* (Jan. 1903), 359.

65 *Annual Report of the President, Queen's University*, 1911–12, 16.

66 Ibid., 16.

67 In his report for the year 1904–5 to the university president, Cox wrote: "It should be mentioned that Dr. T.G. Goldlewski ... was attracted by the growing reputation of the laboratory from Cracow University for a year's work at McGill." *Annual Report of the President, McGill University*, 1904–5, 46; see also ibid., 1902–3, 35.

68 R. Falconer to Premier Ferguson, 20 Sept. 1928, cited by Ross, "Development of the PHD," 315.

69 *Annual Report of the President, Queen's University*, 1918–19, 41.

70 A.B. Macallum to R. Falconer, 29 Aug. 1908, cited by Ross, "Development of the PHD," 270.

71 H.J. Cody, "A Chapter in the Organization of Higher Education in Canada, 1905–1906," *TRSC* (third series), 40 (1946), section II, 98.

72 For further details on the evolution of the movement for industrial research, see Philip Enros, "The University of Toronto and Industrial Research in the Early Twentieth Century," in R. Jarrell and A. Roos, eds., *Critical Issues in the History of Canadian Science, Technology and Medicine*

(Thornhill 1983), 155–66; and Enros, "The Bureau of Scientific and Industrial Research and School of Specific Industries: The Royal Canadian Institute's Attempt at Organizing Industrial Research in Toronto, 1914–1918," *HSTC Bulletin*, 7 no. 1 (Jan. 1983), 14–26.

73 The events surrounding creation of the NRC have been presented by Thistle, *Inner Ring*. The summary here is derived essentially from this source.

74 On the origins of the British research council, see Roy M. McLeod and E. Kay Andrews, "The Origins of the D.S.I.R.: Reflections on Ideas and Men, 1915–1916," *Public Administration*, 48 (1970), 23–48.

75 Thistle, *Inner Ring*, 6.

76 Ibid., 9–12. Kirkpatrick was named later, after pressure from the authorities at Queen's University, who felt excluded from the project.

77 Ibid., 9–11, 69.

78 *Annual Report of the NRC*, 1917–18, 22.

79 Ibid., 23.

80 Ibid., 24.

81 Thistle, *Inner Ring*, 29.

82 Third Conference of Canadian Universities, McGill University, 22 and 23 May 1916, in *National Conference of Canadian Universities*, 23–4.

83 A.B. Macallum to A.S. Mackenzie, 28 Nov. 1918, reproduced in Thistle, *Inner Ring*, 43.

84 "Memorandum regarding the Communications of the Principal and Registrar of Queen's University on the Proposed National Research Institute of Canada," reproduced in ibid., 57.

85 Sixth Conference of Canadian Universities, Ottawa, 23 May, 1919, in *National Conference of Canadian Universities*.

86 Ninth Conference of Canadian Universities, Queen's University, 14–16 June, 1923, in ibid., 69.

87 For further detail on the history of the NRC laboratories, which did not open until 1932, see Thistle, *Inner Ring*, passim.

88 *Annual Report of the President, McGill University*, 1917–18, 73.

89 *Annual Report of the President, Saskatchewan University*, 1916–17, 4.

90 *Annual Report of the President, Queen's University*, 1916–17, 15.

91 Ibid., 17.

92 Cited by Ross, "Development of the PHD," 268.

93 J.C. McLennan to R.A. Falconer, 9 Aug. 1916, cited in Ross, "Development of the PHD," 274.

94 Minutes, 1st Meeting, 4–6 Dec. 1916, Minute no. 21, NRC Archives.

95 Ibid.

96 Thistle, *Inner Ring*, 27.

97 *Annual Report of the NRC*, 1917–18, 38.

98 Ibid., 37.

99 Ibid., 1918–19, 42.

100 These figures are compiled from annual reports of the NRC which include a list of award recipients, their research fields, and the universities where they worked.

101 A.B. Macallum to J.C. McLennan, 22 June 1918, cited in Thistle, *Inner Ring*, 36.

102 Fourth Conference of Canadian Universities, 52–9.

103 Ibid., 53. See also Fifth Conference, 1918, 11; and Eighth Conference, 1922, 52–9.

104 A.B. Macallum to Sir George Foster, 25 Aug. 1917, cited in Thistle, *Inner Ring*, 27.

105 On the situation at McGill, see Frost, *McGill University*, II, 177–81; and Thompson, *Graduate Education in the Sciences in Canada* (Toronto 1963), chap. 4.

106 Report of the Committee on Scientific Research, in *Annual Report of the Principal of Queen's University*, 1916–17, 30.

107 Ibid., 35.

108 A.L. Clark to G.Y. Chown, 23 March 1919, Queen's University Archives (QUA), coll. 2400, Box 2.

109 "Report of Committee on Scientific Research," in *Annual Report of the President, Queen's University*, 1926–27, 47.

110 A.L. Hughes to A.N. Shaw, 25 March 1922, McGill University Archives (MUA), RG 32, Box 47, File 1182. I would like to thank Philip Enros for having brought this letter to my attention.

111 J.A. Gray to E. Rutherford, 27 Oct. 1926, QUA, Gray Papers, Box 4.

112 NRC, Minutes, 5th Meeting, 12–21 May 1917, 4.

113 NRC, Minutes, 3rd Meeting, 13–17 Feb. 1917, 12.

114 Regulations Governing the Award for Grants for Research, *Annual Report of the NRC*, 1917–18, 30.

115 On this issue, Clark frequently deplores the fact that the committee must buy apparatus that, in other universities, are in standard use. See *Report of the Commission of Scientific Research*, in *Annual Report of the President, Queen's University*, 1922–23, 40.

116 *Annual Report of the NRC*, 1932–33, 18.

117 On the impact of the Depression on NRC activities see Wilfrid Eggleston, *National Research in Canada: The NRC 1916–1966* (Toronto 1968), 61–85.

118 I would like to thank Erwin Levold, of the Rockefeller Archives, Tarrytown, New York, for this information.

119 *Annual Report of the NRC*, 1926–7, 17.

120 The annual reports of the NRC do not always mention the amount of money given to each project. However, there is a list containing this

information for the period 1917–37. Though it is not complete, I have used it to make this evaluation. See "National Research Council. List of research grants to individual applicants," National Archives of Canada (NAC), RG 77, vol. 276, file A-A-2-14.

121 H. Blair Neatby, "The Gospel of Research: The Transformation of English-Canadian Universities," *TRSC* (fourth series), 20 (1982), 275.

CHAPTER THREE

1 These numbers include only individual research grants. They exclude applied research contracts given to the associate committees or to other institutions and for which we have no information, though they constitute about 80 per cent of the NRC's research budget. (Compare with Table 2.3.)

2 The information that follows comes from the annual reports of the NRC.

3 See Luc Chartrand, Raymond Duchesne, and Yves Gingras, *Histoire des sciences au Québec* (Montreal 1988), 244–6.

4 Minutes of the Committee on Scientific Research, 19 March 1923, Queen's University Archives (QUA), coll. 1221.

5 Ibid., 30 March 1931.

6 The history of the University of Toronto's physics department has been told by Elizabeth J. Allin, *Physics at the University of Toronto, 1843–1980* (Toronto 1981). For McLennan, see also H.H. Langton, *Sir John Cunningham McLennan: A Memoir* (Toronto 1939). For a synthesis of Toronto's contribution to work in low-temperature physics, see E.F. Burton, *Superconductivity* (Toronto 1934), and H. Grayson Smith and J.O. Wilhelm, "Superconductivity," *Reviews of Modern Physics*, 7 (Oct. 1935), 237–71.

7 For Toronto's contribution to the study of radioactivity, see Lawrence Badash, *Radioactivity in America: Growth and Decay of a Science* (Baltimore 1979).

8 J.C. McLennan to R.A. Falconer, 3 Oct. 1919, University of Toronto Archives (UTA), Falconer Papers.

9 *Report of the President, University of Toronto, 1920–21*, 26–7.

10 J.C. McLennan to R.A. Falconer, 20 Oct. 1923, UTA, Falconer Papers, Box 84.

11 Shrum tells the story of this discovery as well as that of the liquefaction of helium in his autobiography, with Peter Stursberg, Clive Cocking, ed., *Gordon Shrum: An Autobiography* (Vancouver 1986), 38–48.

12 A.S. Eve, "Sir John Cunningham McLennan 1867–1935," *Royal Society of London: Obituary Notices of Fellows*, I, 579–80. J.C. McLennan and G.M. Shrum, "On the Origin of the Auroral Green Line 5577A and Other Spectra Associated with the Aurora Borealis," *Proceedings of the Royal Society of London*, 108 (1925), 501–12. Of all of McLennan's papers, this

one received the most citations during the 1920s – fourteen between 1926 and 1929, according to data in Institute for Scientific Information, *Physics Citation Index 1920–1929* (Philadelphia 1981).

13 K. Mendelssohn, *La recherche du zéro absolu* (Paris 1966), 221.

14 Elizabeth J. Allen, "Physics at the University of Toronto," *Physics in Canada*, 33 no. 2 (1977), 26–31.

15 A.B. Macallum to W.C. Murray, 4 June 1920, reprinted in Thistle, *The Inner Ring: The Early History of the National Research Council* (Toronto 1966) 96.

16 E. Rutherford to J.C. McLennan, 1 Feb. 1929, UTA, McLennan Papers, Scrapbook letters, p. 100.

17 Among the 465 citations received by McLennan between 1920 and 1929 only sixty were self-citations. The 405 others were distributed as follows: 31 per cent from German journals; 30 per cent British, 27 per cent American, 5 per cent French, et 2 per cent Italian, and 1 per cent others; data compiled from *Physics Citation Index 1920–1929*.

18 A.B. Macallum to W.C. Murray, 4 June 1920, reprinted in Thistle, *Inner Ring*, 97.

19 In 1928 and 1929, for example, he received $1,500 to study electrical methods for determining the humidity content of grain and wood.

20 Between 1920 and 1932, the year of his departure, McLennan rarely had less than fifteen graduate students, whereas Burton and Gilchrist never had more than one or two each. As for McTaggart and Satterly, they had no help whatsoever before the end of the 1920s, when each one managed to hire an assistant.

21 J.S. Foster to N. Bohr, 17 Jan. 1935, Bohr Scientific Correspondence, American Institute of Physics, New York.

22 Jerry Thomas, "John Stuart Foster, McGill University, and the Renascence of Nuclear Physics in Montreal, 1935–1950," *HSPS*, 14 (1983), 357–97.

23 "John Stuart Foster, 1890–1964," *PRSC* (fourth series), 3 (1966), 101–5, and *Biographical Memoirs of the Royal Society of London*, 12 (1966), 147–61. Foster presented a synthesis of his work in his presidential address to section III of the Royal Society of Canada in June 1949. See J.S. Foster, "A Quarter-Century of Research in Physics," *TRSC* (third series), 43 (1949), section III, 1–13.

24 For further detail on physics at McGill in this period, see Yves Gingras, "La physique à McGill entre 1920 et 1940: la réception de la mécanique quantique par une communauté scientifique périphérique," *HSTC Bulletin*, 17 (Jan. 1981), 15–39.

25 This overview of the evolution of research in the different physics departments is based on an analysis of the annual reports of the departments and of the NRC. These documents contain each year a summary of projects funded. I have in addition, constituted a list of

master's and doctoral theses with the help of information furnished by
the departments, complemented by other biographical sources, such as
American Men of Science. A short survey of the period 1925–30 may be
found in W.E.K. Middleton, *Physics at the National Research Council,
1929–1952* (Montreal 1979), 3–5. For 1900–32, see A.N. Shaw,
"Canadian Contributions to Physics," *Royal Society of Canada Fifty Years
Retrospect 1882–1932* (Ottawa 1932), 97–105.

26 "Robert William Boyle, 1883–1955," *PRSC* (third series), 49 (1955),
63–6; A.N. Shaw, "Recollections of Robert William Boyle, 1883–1955:
A Distinguished Canadian Physicist," *Physics in Canada,* 10 no. 4 (1955),
21–8.

27 J.L. Heilbron, "Physics at McGill in Rutherford's Time," in M. Bunge and
W. Shea, eds., *Rutherford and Physics at the Turn of the Century* (New York
1979), 50.

28 "Arthur Lewis Clark, 1873–1956," *PRSC* (third series), 52 (1958), 71–5.

29 "John Kenneth Robertson, 1885–1958," *PRSC* (third series), 53 (1959),
123–8.

30 J.C. McLennan and P. Lowe, "On the Structure of Balmer Series Lines of
Hydrogen," *Proceedings of the Royal Society of London,* (Series A), 100
(1921), 217–26.

31 Two students received the NRC bursary in 1927, and three more the
following year.

32 Gordon H.E. Sims, *A History of the Atomic Energy Control Board* (Ottawa
1980), 66–7.

33 "Joseph Alexander Gray, 1884–1966," *PRSC* (fourth series), 6 (1968),
107–18. See also F.W. Gibson, *Queen's University,* II (Kingston 1983),
125–7, 167–9, 265–6; B.W. Sargent, "Nuclear Physics in Canada in the
1930s," in William Shea, ed., *Otto Hahn and the Rise of Nuclear Physics*
(Dordrecht 1983), 221–40.

34 The history of Dalhousie's physics department has been told by J.H.L.
Johnstone, *A Short History of the Physics Department, Dalhousie University
1838–1956* (Halifax 1971).

35 "George Hugh Henderson, 1892–1949," *PRSC* (third series), 44 (1950),
77–82.

36 The history of the University of Saskatchewan's physics department has
been told by B.W. Currie, *The Physics Department, 1910–1976: University of
Saskatchewan* (Saskatoon 1976).

37 For further detail on this point, see Robert H. Kargon, *The Rise of Robert
Millikan* (Ithaca 1982), 153.

38 *Annual Report of the President, Saskatchewan University,* 1935–36, 25.

39 Currie, *Physics Department,* 23.

40 For further detail on the circumstances that brought Herzberg to
Saskatoon, see John Spinks, *Two Blades of Grass: An Autobiography*

(Saskatoon 1980), 35, 43–5. See also Lawrence D. Stokes, "Canada and Academic Refugees from Nazi Germany: The Case of Gerhard Herzberg," *Canadian Historical Review*, 57 (1976), 150–70.

41 *Annual Report of the President, Saskatchewan University*, 1937–38, 24, 1938–39, 27.

42 Gibson, *Queen's University*, 103.

43 Ibid., 448.

44 Allin, *Physics at the University of Toronto*, 25.

45 Currie, *Physics Department*, 35. For the situation at McGill, see Harris, *History of Higher Education*, 353.

46 For an overview of research in francophone universities during this period, see Cyrias Ouellet, *La vie des sciences au Canada français* (Quebec 1964).

47 *Physics in Canada* (1949), 14.

48 Wilfrid Eggleston, *Canada's Nuclear Story* (Toronto 1965), 107.

49 Chartrand, Duchesne and Gingras, *Histoire des sciences*, 420–6.

50 B. Schroeder has analysed the evolution of the origin of the diplomas held by Canadian university professors for the period 1950–1972. In scientific disciplines, the author reports a drop in the proportion of Canadian diplomas: from 60.9 per cent in 1950 to 41.8 per cent in 1972. For more detail, see B. Schroeder-Gudehus, "Les paradoxes de la croissance universitaire," *Revue canadienne de l'enseignement supérieur*, 4 no. 2 (1974), 10–20.

51 J.C. McLennan to R.A. Falconer, 13 Dec. 1926, UTA, Falconer Papers, Box 102.

52 J.C. McLennan to R.A. Falconer, 11 May 1928, UTA, Falconer Papers, Box 108.

53 Falconer explained his decision in a letter to McLennan, 7 May 1930, UTA, Falconer Papers, Box 121.

54 With regard to Synge, see Gilbert de B. Robinson, *The Mathematics Department in the University of Toronto (1827–1978)* (Toronto 1979), 28, 41–5, and Leopold Infeld, *Why I Left Canada* (Montreal 1978).

55 See, for example, N. Bohr to J.S. Foster, 21 Jan. 1935, and Foster to Bohr, 25 Feb. 1935, concerning the interpretation of the Stark effect in a mixture of hydrogen and deuterium; Bohr Scientific Correspondence, American Institute of Physics.

56 P.R. Wallace, "Theoretical Physics in Canada," *Physics in Canada* (1950), 29–31.

57 W.P. Thompson, *Graduate Education in the Sciences in Canada* (Toronto 1963), 17.

58 The 1950s have been covered by M. Christine King in *E.W.R. Steacie and Science in Canada* (Toronto 1989).

CHAPTER FOUR

1 Five papers published between 1895 and 1900, all bearing on the
properties of electrolytes, were published simultaneously in two journals:
one in *TRSC* and in *Physical Review* in the United States and four in the
Transactions of the Nova Scotia Institute of Science and in *Philosophical
Magazine.*

2 The formation of the field of physics in the second half of the 19th
century has not yet been adequately analysed. However, useful elements
may be found in R. Sviedrys, "The Rise of Physics Laboratories in
Britain," *HSPS*, 7 (1976), 405–36, and Robert Silliman, "Fresnel and the
Emergence of Physics as a Discipline," *HSPS*, 4 (1974), 137–62. On the
notion of a scientific field, see Pierre Bourdieu, "Le champ scientifique
et les conditions sociales du progrès de la raison," *Sociologie et sociétés,* 7
no. 1 (1975), 91–118.

3 The literature on Rutherford is abundant. For an analysis of his work at
McGill, see Thaddeus J. Trenn, *The Splitting of the Atom: The History of the
Rutherford-Soddy Collaboration* (London 1977), and A.S. Eve, *Rutherford*
(London 1939).

4 McLennan and Rutherford's work on radioactivity is analysed in the
context of the general scientific productivity of the time by L. Badash,
Radioactivity in America: Growth and Decay of a Science (Baltimore 1979).

5 The difference between Callendar and Barnes's work and that done by
Rutherford is clearly delineated by John L. Heilbron, "Physics at McGill
in Rutherford's Time," in M. Bunge and W. Shea, eds., *Rutherford and
Physics at the Turn of the Century* (New York 1979), 42–73.

6 *PRSC* (second series), 11 (1905), ii.

7 Ibid., ii–iii.

8 Ibid., xiv.

9 A.S. Eve and D. McIntosh, "The Amount of Radium Present in Typical
Rocks in the Immediate Neighbourhood of Montreal," and A.S. Eve,
"On the Amount of Radium Emanation in the Atmosphere near the
Earth Surface," both in *Bulletin of the Royal Society of Canada*, no. 1 (21
June 1907), later published in *TRSC* (third series), 1 (1907), 13–24.

10 J.C. McLennan and W.T. Kennedy, "On the Radioactivity of Potassium
and Other Alkali Metals," *Bulletin of the Royal Society of Canada*, no. 4
(1908), published later in *TRSC* (third series), 2 (1908), 15–30, and
Philosophical Magazine, 16, 93 (Sept. 1908) 377–95.

11 For more detail on this period, see N. Feather, *Lord Rutherford* (London
1973), 100–8.

12 Daniel Kevles, *The Physicists* (New York 1978), 76.

13 *Physical Review*, 2 (1895), 321–43.

14 Before accepting, he reflected on the advantages and inconveniences
of the two positions. He was perhaps influenced by his colleague who,

commenting on the situation, wrote saying, "We're looked on with suspicion from the very fact we've been here so long"; L.M. Keashy to A.S. Mackenzie, 19 March 1905. See also 19 April 1905 and 29 May 1905 and I. Harkness to Mackenzie, 16 Sept. 1904 and 22 May 1905. Dalhousie University Archives (DUA), Mackenzie Collection, MS 2–43.

15 A.S. Mackenzie, "The Deflexion of Alpha-Rays from Radium and Polonium," *Philosophical Magazine* (sixth series), 10 (1905), 538–48.

16 E. Rutherford to A.S. Mackenzie 21 Nov. 1905, DUA, Mackenzie Collection. In this letter, he addressed Mackenzie as "Dear Dr. Mackenzie." In the letters that followed, he wrote "My Dear Mackenzie."

17 E. Rutherford to A.S. Mackenzie, 27 Jan. 1907, ibid.

18 *PRSC* (second series), 5 (1899), vi.

19 E. Rutherford to A.S. Mackenzie, 10 Feb. 1907, DUA, Mackenzie Collection.

20 H.T. Barnes to A.S. Mackenzie, 20 Feb. 1908, ibid. See also the letters of 27 April and 2 June 1908.

21 During the 1920s, this type of negotiation continued. At the beginning of 1921, R.F. Ruttan, of McGill's chemistry department, complained that "the balance of the different interests in the Section" had not been maintained; R.F. Ruttan to J. Patterson, 11 Jan. 1921. See also J. Patterson to A.L. Clark, 9 Feb. 1921; R.F. Ruttan to J. Patterson, 30 Dec. 1921; and J. Patterson to L.V. King, 30 Dec. 1922. This correspondence may be found in John Patterson's papers; from 1921 to 1935, he was secretary of section III. National Archives of Canada (NAC), John Patterson Papers, RG 93, vol. 83, File "Royal Society, 1920–1925."

22 In a study of the power structure of the Canadian élite, John Porter remarked that the Royal Society was not constituted by the élite of Canadian scientists. However, he did not attempt to explain why this was the case. See John Porter, *The Vertical Mosaïc* (Toronto 1965), 496–7.

23 They were: H.T. Barnes, elected in 1911, J.C. McLennan (1915), A.S. Eve (1917), L.V. King (1924), J.A. Gray (1932), J.S. Foster (1935), and G.H. Henderson (1941).

24 This information is taken from the minutes of the annual meetings of the APS, published since 1903 in *Physical Review*, and from *Bulletin of the American Physical Society*, 24 (1930), xxx.

25 In 1930, for example, the presence of eighteen visitors was recorded. *PRSC* (third series), 24 (1930), xxx.

26 Minutes, 23rd annual meeting, 30 Dec. 1921, Toronto, *Physical Review*, 19 (Jan.–June 1922), 374–79.

27 *PRSC* (third series), 17 (1923), xxxii; 18 (1924), xxxiv; 19 (1925), xxiv.

28 Ibid., 21 (1927), xxxviii; 23 (1929), xxix.

29 Ibid., 15 (1921), xxii. Members returned to this project at more or less regular intervals, without, however, ever making any modifications. See, for example, ibid., 20 (1926), xxxvi–xxix, and 31 (1937), xxx.

30 L.V. King to J. Patterson, 13 Dec. 1922, NAC, Patterson Papers, File "Royal Society, 1920–1925."

31 *PRSC* (third series), 17 (1923), xlviii.

32 Ibid., 18 (1924), xli.

33 Kevles, *The Physicists*, 41–4, 76–7.

34 *PRSC* (third series), 2 (1909), vi.

35 Ibid., 7 (1913), xxv–xxvii.

36 Ibid., 8 (1914), xv.

37 Ibid., xvi.

38 Ibid., 12 (1918), xxx.

39 *PNRC*, 21st Meeting, 21–22 March. 1919, 14.

40 Ibid.

41 Ibid., 22nd Meeting, 2–3 May, 1919, 3.

42 Ibid., 25th Meeting, 19 Sept., 1919, 9.

43 *PRSC* (third series), 14 (1920), ix.

44 Ibid., 13 (1919), xxvi.

45 NRC Associate Committee on Physics and Engineering Physics, Minutes, 3rd Meeting, 17–21 May, 1921, 4, NRC Archives.

46 *PRSC* (third series), 15 (1921), xxi.

47 *PNRC*, 40th Meeting, 20–21 May, 1921, 20.

48 Ibid.

49 NRC Associate Committee on Physics and Engineering Physics, Minutes, 5th Meeting, 16 May 1922, 2.

50 Ibid., 11, Appendix E, "Committee on Publication of Scientific Papers in Canada."

51 Ibid.

52 *PRSC* (third series), 16 (1922), xxix.

53 *National Conference of Canadian Universities* (n.p., n.d.), Ninth Conference, 14–16 June 1923, 14; Tenth Conference, 2–4 June 1925, 95–6.

54 J. Patterson to J.A. Gray, 21 Nov. 1924, QUA, J.A. Gray Collection, coll. 1057, Box 1.

55 *PNRC*, 61st Meeting, 7 Jan. 1925, 3. Tory was named honorary director of the NRC on 5 October 1923. See Mel Thistle, *The Inner Ring: The Early History of the National Research Council* (Toronto 1966), 123.

56 *PNRC*, 65th Meeting, 17 Dec. 1925, 5. The paper written by McLennan and his student, A.B. Mclay, was entitled "On the Structure of the Arc Spectrum of Manganese," *TRSC* (third series), 20 (1926), section III, 89–120.

57 *PRSC* (third series), 20 (1926), xxxi.

58 Ibid., 21 (1927), xxvii.

CHAPTER FIVE

1 *PNRC*, 53rd Meeting, 25–26 May, 1923.
2 NRC Associate Committee on Physics and Engineering Physics, 6th Meeting, 24–26 May 1923, 8–9.
3 H.M. Tory to W.C. Murray, 3 Feb. 1925, reproduced in Mel Thistle, *Inner Ring: The Early History of the National Research Council* (Toronto 1966), 173.
4 *PNRC*, 68th Meeting, 7–9 June, 1926, 17.
5 Members of the committee were J. Patterson, A.S. Eve, R.W. Boyle, G.H. Henderson, H.M. Tory, and Augustin Frigon (of the Ecole Polytechnique in Montreal); ibid., 18.
6 See the financial statements published yearly in *PRSC*.
7 *PNRC*, 74th Meeting, 20 Jan. 1928; 77th Meeting, 25–26 Oct. 1928; 78th Meeting, 1–2 Feb. 1929; 80th Meeting, 8 July 1929.
8 Ibid., 78th Meeting.
9 "Foreword," *Canadian Journal of Research* (*CJR*), 1 no. 1 (May 1929), 3.
10 *Report of the President, NRC*, 1928–29, 11.
11 Terry Shinn, for example, has taken as evidence for the quality of French mathematics the fact that French mathematicians had little difficulty publishing abroad; "The French Science Faculty System," *HSPS*, 10 (1979), 298. Those who analyse national scientific communities from the point of view of dependency theory often implicitly use foreign publication as an indicator of dependence. See, for example, M. Fournier et al. "Le champ scientifique québécois, structure, fonctionnement et fonctions," *Sociologie et sociétés*, 7 no. 1 (1975), 119–32; Michel Leclerc, *La science politique au Québec* (Montreal 1982).
12 In his study of American physicists, Daniel Kevles defines productive physicists as those who published at least one article every three years during the period 1894–1915. My calculation, in contrast, is based on the total number of articles produced by Canadian physicists. This difference may result in an overestimation of the number of American papers published abroad, as less productive physicists probably had a tendency to publish more at a local level. See Daniel Kevles, "The Physics, Mathematics and Chemistry Communities: A Comparative Analysis," in A. Oleson and J. Voss, eds., *The Organization of Knowledge in Modern America, 1860–1920* (Baltimore 1979), 139–72.
13 The mean number of citations received by article cited between 1920 and 1929 was 2.5 (including self-citations), according to a random sample of 250 articles cited in Institute for Scientific Information, *Physics Citation Index 1920–1929* (Philadelphia 1981). The mean falls to 2.0 if we exclude citations attributed by researchers of the same nationality as the author cited, on the hypothesis that the proportion of national citations

is similar to that of Canada.

14 The distribution by country of citations accorded Indian physicists is more or less identical to that accorded Canadians. The proportion of articles cited that were published in India was less than half of what it should have been according to the laws of chance: between 1900 and 1929, Indian physicists published 26 per cent of their work in Indian journals for which they received only 10 per cent of the citations. Australian physicists received only 1 per cent of their citations for work published in Australian journals, where they published 19 per cent of their papers. Moreover, the mean number of citations per article cited was only 1.8 for Indian physicists and 2.6 for Australians. The small number of Australian physicists explains this figure, which is far from the mean. Taken together, Indians and Australians received around two citations per article cited, which was the mean for the population as a whole. Citations were compiled on the basis of the *Physics Citation Index 1920–1929*, I and II. Data concerning number of articles published by Indian and Australian researchers come from L. Pyenson and M. Singh, "Physics on the Periphery: A World Survey, 1920–1929," *Scientometrics*, 6 no. 5 (1984), 279–306.

15 *PNRC*, 84th Meeting, 29 March, 1930, 3.

16 *Annual Report of the NRC*, 1932–33, 14; 1939–40, 63.

17 Ibid., 1945–46, 60.

18 Ibid., 1928–29, 11.

19 In an article commemorating the fiftieth anniversary of the *CJR*, N.T. Gridgeman states that although there was often conflict between the *CJR* and the *Proceedings*, there is "no evidence of an ill effect on the *Transactions*." It is hard to believe that the disappearance of almost all the scientific articles from the *Transactions* was in the interests of the directors of the Royal Society or that this disappearance went unnoticed. See N.T. Gridgeman, "A Semicentennial: The NRCC Research Journals, 1929–1979," *Canadian Journal of Physics*, 57 (July 1979), vi.

20 *Annual Report of the NRC*, 1935–36, 69.

21 These numbers represent the mean for all research areas. The distribution by country may vary according to specialty.

22 Gridgeman, "A Semicentennial," xi.

23 See *PRSC* (third series), 39 (1945), 35–6, and *Annual Report of the NRC*, 1945–46, 58–9. On the origin of chemists' associations, see T.H.G. Michael, "The Association Aspects of Chemical Engineering in Canada," in William F. Fuller, ed., *History of Chemical Engineering: Advances in Chemistry*, no. 190 (Washington 1980), 199–204.

24 *PNRC*, 121st Meeting, 18 March 1938, 6.

25 *PRSC*, (third series), 34 (1940), 48.

26 Ibid., 35 (1941), 56.

27 "Minutes of the Council of the CAP," 2 June 1951, CAP Archives, Ottawa, Box 1, file 1.5

CHAPTER SIX

1 Russell Moseley, "Tadpoles and Frogs: Some Aspects of the Professionalization of British Physics, 1870–1939," *Social Studies of Science,* 7 (1977), 433.

2 Not only is there an abundant literature on the "professionalization" of the sciences, but many studies use this term implicitly without giving it a precise meaning. On the nineteenth century, see Everett Mendelsohn, "The Emergence of Science as a Profession in Nineteenth-Century Europe," in K. Hall, ed., *Management of Scientists* (Boston 1963), 3–48; R. Steven Turner, "The Growth of Professional Research in Prussia, 1815 to 1848 – Causes and Context," *HSPS,* 3 (1971), 137–82. See also N. Reingold, "Definitions and Speculations: The Professionalization of Science in America in the Nineteenth Century," in A. Oleson and S.C. Brown, eds., *The Pursuit of Knowledge in the Early American Republic* (Baltimore 1976), 33–9; Robert Fox, "La professionnalisation: un concept pour l'historien de la science française au XIX siècle," *History and Technology,* 4 (1987), 413–22. For an excellent critique of the use of the concept of professionalization, see Dorinda Outram, "Politics and Vocation: French Science, 1793–1830," *British Journal for the History of Science,* 13 (1980), 27–43.

Although Charles Rosenberg had as early as 1979 suggested distinguishing "discipline" and "profession," few historians have followed his advice. See Charles Rosenberg, "Toward an Ecology of Knowledge: On Discipline, Context, and History," in A. Oleson et J. Voss, eds., *The Organization of Knowledge in Modern America, 1860–1920* (Baltimore 1979), 440–55. For examples of the "impressionistic" use of the notion of professionalization in the history of physics, see Daniel Kevles, *The Physicists* (New York 1978), and R.W. Home, "Between Classroom and Industrial Laboratory: The Emergence of Physics as a Profession in Australia," *Australian Physicist,* 20 (Aug. 1983), 163–7, and, "Origins of the Australian Physics Community," *Historical Journal,* 20 (1982–83), 383–400. Maximum confusion was no doubt attained by Roger L. Geiger, who, in an otherwise useful book, speaks of the "professionalization of disciplines" without defining either term. See Roger L. Geiger, *To Advance Knowledge: The Growth of American Research Universities, 1900–1940* (New York 1986), 21–2.

3 Terence Johnson, *Professions and Power* (London 1972); Magali Sarfatti-Larson, *The Rise of Professionalism: A Sociological Analysis* (Berkeley 1977).

4 Sarfatti-Larson, *The Rise of Professionalism,* 33.

5 The literature on scientific disciplines is also abundant. See, for example, G. Lemaine et al., *Perspectives on the Emergence of Scientific Disciplines* (Chicago 1976). For a complete bibliography, see S*tudies of Scientific Disciplines: An Annotated Bibliography*, National Science Foundation, NSF 83–7, Washington, 1982.

6 *PNRC*, 14th Meeting, 25 May 1917, 4; 18th Meeting, 1–2 Nov. 1918, Appendix B, NRC Archives, Ottawa.

7 Moseley, "Tadpoles and Frogs," 434, and Roy Mcleod and Kay Andrews, "Scientific Advice in the War at Sea: The Board of Invention and Research," *Journal of Contemporary History*, 6 (1971), 3–40.

8 The authors of these comments were E.F. Burton of Toronto, A.L. Clark of Queen's, H.F. Dawes of McMaster, R.F. Haley of Acadia, H.L. Hogg of Saskatchewan, L.V. King and A.N. Shaw of McGill, W.C. Baker of Western Ontario, and P.J. Nicholson of St-Francis Xavier.

9 *PNRC*, 18th Meeting, 1–2 Nov. 1918, Appendix B, "Conference of Physicists," 3.

10 Ibid.

11 *PNRC*, 27th Meeting, 12–13 Dec. 1919, 6. H.H. Langton, *Sir John Cunningham McLennan: A Memoir* (Toronto 1939).

12 *Annual Report of the NRC*, 1920–21, 42.

13 To give but one example, the report of the committee for the year 1930–31 noted that "other subjects discussed were the development of altimeters, radio standards, electrical methods of moisture determination, colloids, artificial grounds, insulating materials, corrosion alloys, cathode rays, clay conduits, scholarships for physicists and the training of the physicists in the universities." All this during a two-day meeting at the physics department at the University of Toronto. Ibid., 1930–31, 89.

14 In addition to hearing McLennan, who presented his work on atomic spectroscopy, and Burton, who spoke on the properties of colloidal solutions, physicists present at the colloquium listened to a series of seminars by Ludwick Silberstein on theories of relativity and Irving Langmuir on atomic structure.

15 A motion voted at the end of the meeting thanked the NRC "for making it possible for many of the physicists to attend." NRC Associate Committee on Physics and Engineering Physics, Minutes, 2nd Meeting, 14–15 Jan. 1921, 13.

16 Ibid., 11.

17 Ibid., 11.

18 Ibid., 12, my emphasis.

19 During their meeting, the members of the committee frequently adopted "resolutions" of a collective nature. For instance: "For a perfectly co-ordinated development of industry in Canada, there should

be available ... specialists in physics and engineering physics ... who have received a training in research work in these two fields," or "The chief need of physicists in Canada at present is a Canadian Journal that will enable them to publish scientific papers promptly." *Annual Report of the NRC,* 1920–21, 43–4.

20 Patrick Champagne, "La manifestation: la production de l'événement politique," *Actes de la recherche en sciences sociales,* no. 52–3 (June 1984), 39; see also Luc Boltanski, *Les cadres: la formation d'un groupe social* (Paris 1982), 232–5.

21 In his answer to Mackenzie about creating an association of physicists, Louis King wrote: "[It is] unfortunate that it is not recognized that Engineering is only applied physics ... In order to do away with this distinction ... it would be necessary to revise college curricula." Though King could not *by himself* modify the relations between scientific disciplines, an organization of physicists could try to do so by speaking in a collective manner; *PNRC,* 18th Meeting, 1–2 Nov. 1918, "Conference on Physics," 3.

22 NRC Associate Committee, Minutes, 3rd Meeting, 17–18 May 1921, 11. The secretary wrote not "Canadian physicists" but "physicists of Canada." In the same way, the report of activities of the committee for the year 1920–21 spoke of "physicists in Canada," as if physicists had not yet achieved a national identity.

23 Ibid., 11.

24 Ibid., Minutes, 5th Meeting, 16 May 1922, 2.

25 As A. Norman Shaw wrote to A.S. Mackenzie in 1918, associations of physicists "have usually been formed for affording channels of communication for ideas and investigations already in mature action, rather than stimulating interests still in their infancy." According to him: "Stimulation in research will come from public recognition of research, the backing of the government, scholarships, etc." *PNRC,* 18th Meeting, 1–2 Nov. 1918, "Conference of Physicists."

26 NRC Associate Committee, Minutes, 1st Meeting, 21 May 1920, 4.

27 Ibid., Minutes, 4th Meeting, 31 Dec. 1921, 6.

28 Ibid., Minutes, 13th Meeting, 17–18 Oct. 1930, 9–10. McLennan and Eve had just been named dean of graduate studies at Toronto and McGill, respectively.

29 Never attained, this objective was also that of the NCCU. See G. Pilkington, "A History of the National Conference of Canadian Universities," DPHIL thesis, University of Toronto, 1974, 19.

30 *Annual Report of the NRC,* 1921–22, 30; my emphasis.

31 NRC Associate Committee, Minutes, 2nd Meeting, 14–15 Jan. 1921, 10.

32 Ibid., Minutes 13th Meeting, 17–18 Oct. 1930, 12.

33 Ibid.

34 Ibid.
35 The list of the committee's members is given for the last time in the annual report of the NRC for the year 1931–32 (p. 157).
36 *Annual Report of the NRC,* 1928–29, 49.

CHAPTER SEVEN

1 For more details, see *Survey of Industrial and Scientific Laboratories in Canada,* Dominion Bureau of Statistics, Ottawa, 1941. On the history of provincial research councils, see F. Anderson, O. Berseneff-Ferry, and P. Dufour, "Le développement des Conseils de recherche provinciaux : quelques problématiques historiques," *HSTC Bulletin,* 7 no. 1 (Jan. 1983), 27–44.
2 On the NRC at the end of the war, see *Annual Report of the NRC,* 1946, 18–24.
3 Roy McLeod and Kay McLeod, "The Contradictions of Professionalism: Scientists, Trade Unionism and the First World War," *Social Studies of Science,* 9 (1979), 1.
4 The information that follows is derived from the *Bulletin of the Montreal Branch: Canadian Association of Scientific Workers,* no. 2 (Jan. 1945), and Peter Keating, "Engineers, Scientists, and Collective Bargaining: The Rise and Decline of the Canadian Council of Professional Engineers and Scientists (1944–1949)," Unpublished, Montreal, 1982.
5 Needless to say, the order-in-council was prompted not by scientists' activities but by workers organized into a number of unions. For more detail, see L.S. Macdowell, "The Formation of the Canadian Industrial Relations System during World War Two," *Labour/Le Travailleur,* 3 (1978), 175–96.
6 J.D. Bernal, *The Social Function of Science* (London 1939), 400.
7 For a review of CASCW activities, see Paul Dufour, "Les "Eggheads" et l'espionnage: les relations des scientifiques américains, canadiens et britanniques à l'affaire Gouzenko de 1946," mémoire de maîtrise, Université de Montréal, 1979, 59–66.
8 Cited by P. Keating, "Engineers, Scientists," 12.
9 Montreal Branch, CASCW, *Newsletter,* no. 6 (1 Jan. 1949), 4.
10 *Bulletin of the Canadian Association of Physicists* (BCAP), 5 no. 2 (Dec. 1949), 2. In 1978, A.D. Misener, one of the founders of CAPP, published two articles on the history of the association. See *Physics in Canada,* 34 no. 5 (Sept. 1978), 103–6; 35 no. 5 (Sept. 1979), 115–18. My presentation is based on a detailed analysis of the CAP archives, deposited at the National Archives of Canada (NAC).
11 Fred E. Coombs to Lt. Col. Eric Phillips, 9 July 1945, NAC, CAP Papers, MG 28 I 289, vol. 3, File "Corres. n.d. 1945." Created in the summer of 1940, Research Enterprises Ltd was a crown corporation specializing in the

production of scientific instruments for military use. It closed in 1945. For the history of this company, see Robert Bothwell and William Kilbourn, *C.D. Howe: A Biography* (Toronto 1979), 206; W. Eggleston, *Canada's Nuclear Story* (Toronto 1965), 169; and David Zimmerman, "Radar and Research Enterprises Limited: A Study of Wartime Industrial Failure," *Ontario History*, 80 no. 2 (1988), 121–42.

12 *Quarterly Bulletin of the Canadian Association of Professional Physicists*, 1 no. 1 (1 July 1945), 1.

13 NAC, CAP Papers, MG 28 I 289, 79/408, vol. 1, file "Minutes 1945–1950."

14 CAP Papers, vol. 1, file 11, undated constitution, surely prepared in 1945 or in 1946.

15 J.P. Sandiford to F.E. Coombs, 22 April 1945, CAP Papers, 79/408, vol. 3, file "corres. n.d., 1945."

16 Ibid., R.D. Dearle to F.E. Coombs, 21 May 1945. A flyer published in 1948, when university physicists had gained a majority within the association, explained the objectives of CAP, noting that there had been "some misunderstanding of, and even some opposition to the association which was also due to the fact that the initiators were relatively unknown to the majority of Canadian physicists"; *Canadian Association of Physicists: Purpose and History* (Toronto 1948), 4.

17 The term "professional" in clause 3 of Allen's charter may give the misleading impression that both groups had the same objective. To industrial physicists, "professional" meant occupational control. To Allen and university physicists in general, it referred, in a somewhat general and abstract way, to the quality of physicists' training. R.J. Allen to F.E. Coombs, 15 June 1945, and R.C. Dearle to Coombs, 21 May 1945, CAP Papers, 79/408, vol. 3, file "corres. n.d. 1945."

18 Ibid., F.E. Coombs to R.J. Allen, 26 June 1945.

19 Ibid., F.E. Coombs to P.J. Sandiford, 29 April 1945.

20 Ibid., Alex J. Ferguson to N.J. Abbott, 10 Dec. 1945.

21 Ibid., N.J. Abbott to A.J. Ferguson, 6 Jan. 1946.

22 J.J. Brown, "Physics in Canada's Industrial Reconversion Programme," *Industrial Canada*, 46 no. 4 (Aug. 1945), 69–71.

23 Ibid.

24 R.W. Boyle to J.O. Wilhelm, 17 Feb. 1945, cap Papers, vol. 2, file 2.1.

25 A.N. Shaw, "Recollections of Robert William Boyle 1883–1955: A Distinguished Canadian Physicist," *Physics in Canada*, 10 no. 4 (1955), 27.

26 E.J. Allin, "To the Members of the Council of C.A.P.," 19 Oct. 1949, CAP Papers, vol. 1, dossier 1.1.

27 CAP Papers, vol. 1, file 1.1, "Constitution and By-laws."

28 BCAP, 7 no. 1 (Nov. 1951), 4.

29 E.J. Allen, "To the Members of the Council of C.A.P.," 19 Oct. 1949, CAP Papers, vol. 1, file 1.1.

30 BCAP, 3 no. 1 (March 1947), p. 1. Of the seventy-seven opponents to the

name CAPP, twenty-nine suggested CAP and twenty-eight, the Canadian Institute of Physics.

31 Among those who answered the questionnaire, ninety-one suggested adding the following objective to the association's charter: "To advance the science of Physics." The author of the report concluded that "otherwise the present objects are approved in principle by a large majority." Ibid., 1.

32 Because many volumes of BCAP were produced on typewritten sheets, it has not been possible to put together a complete collection. Even the present office of the association in Ottawa does not have a complete set. Thus, I have not been able to find the December 1950 issue containing Barnes's letter. However, the essence of its content can be reconstituted on the basis of the replies printed in the following issue. The copies of BCAP that I consulted are to be found at the Ottawa office of CAP.

33 P.J. Sandiford to the editor-in-chief, BCAP, 6 no. 3 (Feb. 1951), 1.

34 Ibid., 4–5.

35 "Some of the Arguments *for* and *against* the Formation of an Ontario Association of Physicists," undated, 1, CAP office, Ottawa. The document was probably written in January 1951. See "The Formation of an Ontario Association of Physicists," BCAP, 7 no. 2 (Feb. 1952), 7–9.

36 Minutes of the Executive Meeting, 27 May 1952. See also ibid., 27 Sept. 1952. CAP Papers, vol. 1, file 1.1.

37 "The Professional Scientist in Canada," Discussion Document, Prepared by the Committee on Professionalism of the Canadian Association of Physicists, Oct. 1987, 14.

CHAPTER EIGHT

1 Minutes, 1st Meeting, Organizing Committee of CAPP, 7 Sept. 1945. NAC, CAP Papers, 70/408, vol. 1, file 1.1.

2 C.J. Mackenzie to N.J. Abbott, 11 Sept. 1945; see also Abbott to Mackenzie, 7 Sept. 1945. CAP Papers, 79/408, vol. 3, file "corres. n.d. 1945."

3 For further detail on this episode, see Eggleston, *Canada's Nuclear Story* (Toronto 1965).

4 In a letter to N.J. Abbott, G.A. Woonton, who was in charge of inviting Foster, wrote: "It is an interesting sidelight on the state of McGill's nerves that Dr. Foster refused to talk to me on the phone until I assured him that CAPP had nothing to do with Russian spies. Has it?" Woonton to Abbott, 2 April 1946. See also Woonton to Wilhelm, 28 March 1946, CAP Papers, vol. 2, file 2.1.

5 *Bulletin of the CAPP*, 11 no. 2 (April 1946), and 11 no. 3 (Nov. 1946). A summary of Watson's presentation appeared in the March, 1947 issue of the *Bulletin*, 4.

6 BCAP, 3 no. 2 (Sept. 1947), 1. This is the first bulletin published under the association's new name.

7 Ibid., 6 no. 3 (Feb. 1951), 5.

8 D.C. Rose, "The Presidential Address, C.A.P., 1950," ibid., 5 no. 5 (July 1950), 4.

9 Ibid., 7 no. 2 (Feb. 1952), 4.

10 *Physics in Canada*, 8 no. 4 (summer 1953), 42.

11 Each annual volume of *Physics in Canada* contains a review of research carried out in Canadian universities, and, beginning with the first issue of 1949, the reports from the Université de Montréal and Laval were written in French.

12 Minutes, Meeting of the Executive, 14 Feb. 1953, 27 May 1953, 29 Jan. 1955, CAP Papers, vol. 1, file 1.2.

13 *TRSC* (fourth series), 1 (1942), 36.

14 "The Award for Achievement in Physics," *Physics in Canada*, 12 no. 3 (autumn 1956), 17.

15 On the occasion of his receiving the Nobel Prize, *Physics in Canada* published a special issue (April 1972) in his honour.

16 Ibid., 14 no. 3 (autumn 1958), 5.

17 Ibid., 15 no. 3 (autumn 1959), 5; 16 no. 3 (autumn 1960), 6; and 17 no. 3 (autumn 1961), 5.

18 Ibid., 9 no. 2 (winter 1953), 6.

19 Ibid., 9 no. 3 (spring 1954), 34.

20 Ibid., 11 no. 2 (winter 1955), 30–7; 13 no. 1 (spring 1957), 36; and 13 no. 4 (spring 1957), 30.

21 Minutes, General Business Meeting, 21 June 1955 and 15 June 1956, CAP Papers, vol. 1, file 1.5. Seven thousand copies of the brochure were printed in English and 3,000 in French.

22 *Physics in Canada: A Career, A Vocation* (n.p., n.d.), probably 1956.

23 *Physics in Canada*, 13 no. 2 (summer 1957), 27.

24 Minutes, Annual General Meeting, McMaster University, 17 June 1958, 2, CAP Papers, vol. 1, file 1.5.

25 *Physics in Canada*, 15 no. 3 (autumn 1959), 31; 17 no. 4 (autumn 1961), 28.

26 For the objectives of the PSSC, see *Physics Today*, 10 no. 3 (March 1957), 28–9. For its application in Canada, see L. Tomaschuk, "The PSSC Course," ibid., 16 no. 2 (summer 1960), 28–31. In ibid., see also 14 no. 3 (autumn 1958), 36, and "The Physical Sciences Study Group," 13 no. 4 (winter 1957), 19–21.

27 The increased role of scientists' representatives in the era of big science has been studied by Daniel Greenberg, *The Politics of Pure Science* (New York 1967).

28 For more detail on the activities at Chalk River, see Eggleston, *Canada's Nuclear Story*.

29   Gordon H.E. Sims, *A History of the Atomic Energy Control Board* (Ottawa 1981), 66.
30   "A High Energy Laboratory for Canada: A Resolution Presented by B.W. Sargent to the Annual Meeting, June 14, 1957," *Physics in Canada*, 13 no. 3 (autumn 1957), 23.
31   In a special issue of *Physics in Canada* published in the fall of 1959 (vol. 15, special issue), CAP presented a history of the project, followed by the dossier that had been submitted to the government. My presentation is based on these texts together with material from the CAP archives.
32   The letters from the department heads are conserved in the CAP Papers, vol. 5, file 5.9.
33   "A High Energy Laboratory for Canada: A Brief from the Canadian Association of Physicists," Hamilton, Sept. 1957, 7, CAP Papers, vol. 5, file 5.9.
34   Ibid., 5. A second version of the report, dated September 1958, offered slightly different information. Costs had risen to $25 million, the construction period to six years, and the energy of the accelerator to 15 GeV. See "A High Energy Laboratory for Canada: A Brief Submitted to the Government of Canada by the Canadian Association of Physicists, September 1958," *Physics in Canada*, 15, special issue (1959), 16–18.
35   "A High Energy Laboratory," Sept. 1957, 27–8.
36   Bruce G. Doern, *Science and Politics in Canada* (Montreal 1972), 107.
37   "A High Energy Laboratory," Sept. 1958 14.
38   Minutes, Annual General Meeting, CAP, Saskatoon, 5 June 1959, 1, CAP Papers, vol. 1, file 1.5.
39   Ibid., 2.
40   Ibid., 2.
41   "Letter to the President," *Physics in Canada*, 15 no. 2 (summer 1959), 2.
42   On TRIUMF, see J.B. Warren, "TRIUMF," ibid., 22 no. 4 (winter 1966), 24–9. The accelerator started operation in 1974; see ibid., 31 no. 4 (June 1975).
43   For a sociological interpretation of "groups" and "parties," see P. Bourdieu, "La représentation politique," *Actes de la recherche en sciences sociales*, no. 36/37 (Feb./March 1981), 3–24, and "La délégation et le fétichisme politique," ibid., no. 52/53 (June 1984), 49–55. See also Luc Boltanski, *Les cadres : la formation d'un groupe social* (Paris 1982).
44   Minutes, Annual General Meeting, CAP, Saskatoon, 5 June 1959, 1, CAP Papers, vol. 1, file 1.5.
45   For more detail on the development of these organizations, see Doern, *Science and Politics in Canada*, and Louise Dandurand, "The Nature of the Politicization of Basic Science in Canada: NRC's Role, 1945–1976," PHD, thesis, University of Toronto, 1982.
46   In 1967, for example, the Science Secretariat asked CAP to prepare a study on the state of physics research in Canada, which allowed the

association to define its priorities and to make them known to the government. See D.C. Rose, *Physics in Canada: Survey and Outlook,* Science Secretariat, Special Study No. 2 (Ottawa, 1967).

CONCLUSION

1 Luc Boltanski, *Les cadres : la formation d'un groupe social* (Paris 1982), 58.
2 This mode of representation may, of course, be modified in the future to respond to changes in the membership of the association and especially the socio-economic conjuncture, which could give greater weight to the "applied" side of the discipline (represented mainly by the industrial milieu).

Index